Science, Nonscience, and Nonsense

Science,
Nonscience,
and
Nonsense

ॐ

Approaching
Environmental
Literacy

Michael Zimmerman

The Johns Hopkins
University Press
Baltimore and London

© 1995 The Johns Hopkins University Press
All rights reserved. Published 1995
Printed in the United States of America on acid-free paper
04 03 02 01 00 99 98 97 96   5 4 3 2 1

The Johns Hopkins University Press
2715 North Charles Street
Baltimore, Maryland 21218-4319
The Johns Hopkins Press Ltd., London

Library of Congress Cataloging-in-Publication Data

Zimmerman, Michael, 1953–
    Science, nonscience and and nonsense: approaching environmental literacy /
    Michael Zimmerman.
        p.   cm.
    Includes index.
    ISBN 0-8018-5090-8
    1. Science—Popular works.   2. Science—Social aspects—Popular works.
    3. Environmental education—Popular works.   I. Title.
    Q162.Z48   1995
    303.48'3—dc20   95-5006

A catalog record for this book is available from the British Library.

For Alex and Jess
and the world they might inherit

It is a capital mistake to theorize before one has data.

—Arthur Conan Doyle, *The Memoirs of Sherlock Holmes*

"Are you lost, daddy?" I asked tenderly.
"Shut up," he explained.

—Ring Lardner, *The Young Immigrunts*

க

Contents

◌⁊

Preface

Since 1959, when C. P. Snow so eloquently explored the gulf between an understanding of the sciences and of the humanities in his classic book *The Two Cultures and the Scientific Revolution,* the problem he described has grown worse. Increasingly large portions of the American public are scientifically illiterate. By this I mean not that many people are unaware of specific scientific "facts," although there can be no doubt that they are. Rather, in pointing to scientific illiteracy I mean something far more rudimentary: an inability to differentiate science from pseudoscience. Science, based on skepticism and dependent on both falsifiability and experimentation, is dramatically different from pseudoscience, based on faith and dependent on gullibility. While pseudoscience can surely be entertaining, whether it be matching astrological signs with a date or playing paranormal parlor games, the situation becomes very different when the inability to distinguish between the two means that important decisions are made on the basis of superstitious drivel. When Nancy Reagan's astrologer advises the president on propitious moments to hold important international meetings, when well-meaning individuals burn cow manure, ghee, rice, and sandalwood in an inverted copper pyramid to rid the planet of air pollution, and when creationists demand that public school science classes teach that the world was created a mere six thousand years ago, the situation is far from entertaining.

Ironically, an acceptance of, and even a reliance on, pseudoscience has burgeoned at a time when science and technology are assuming an ever increasing presence in our daily lives. Technologically we are now capable of things that were not even dreamed of a mere generation ago, but most of us are completely and complacently ignorant of how the majority of our most common technological devices work. In our ignorance we accept their existence and assume that they will do what we expect them to do, as if by a combination of blind faith and magic. At some level such ignorance is perfectly reasonable; none of us should even hope to be equally conversant in nuclear physics, electrical engineering, and biotechnology.

But technology has delivered us to a new plateau, a plateau whose height and vistas have granted us the ability to influence and disrupt our surrounding planet in ways previously unimagined. Marshall McLuhan pointed out two decades ago that through advances in communications the people of the earth have become part of a global village, yet we are only now becoming painfully aware that, with respect to pollution and environmental degradation, our planet is not only small but shrinking. Our technology has enabled us to disrupt the ozone layer, change the composition of atmospheric gases, and produce acid rain on such a scale that whole ecosystems have been destroyed. Indeed, it is possible that our technology is beginning to change the face of our planet in such a way that its very hospitality to humans is threatened.

If we are to deal effectively with current environmental crises and avert future ones, and if we are to put technology to work on our behalf instead of becoming its victims, then we, as members of a democratic society, must soon make a large number of complex decisions. We have to modify our profligate energy habits, face the threats as well as benefits posed by dramatic advances in genetic engineering, and confront the fact that we are rapidly running out of clean drinking water, to name just a few issues. There are three directions in which we might proceed:

— We can simply say that the problems are far too sophisticated for resolution by any participatory democracy and therefore abdicate all responsibility to scientists.
— We can allow a scientifically illiterate public to determine public policy based on "new age" mysticism, cosmic power, or any other pseudoscientific fad that happens along.
— Or we can strive to produce a scientifically educated citizenry.

As a scientist, I recognize that the first option would be very unwise. Even though solutions to our problems might well be based in science and technology, their implementation must be thoroughly grounded in economics, politics, and ethics. Science and scientists will never be our sole saviors.

The second option might well be an accurate description of American society today when you consider that only half of the federal legislators responding to a survey of mine decisively denied the possibility of communicating with the dead. It seems clear that as the scale of our environmental mistakes increases, we can ill afford to continue to bumble along as we have been doing.

The third option is thus the only viable alternative. It is not my wish that every child aspire to be a scientist but that each become as conversant with basic scientific principles and methodology as with literature and sports. Achieving such a goal will not be easy, but we have no realistic alternative. And once our schools start turning out scientifically literate individuals, they will be much more likely to be producing environmentalists at the same time, for it is difficult to be the former without being the latter. As we begin to discover the beauty of nature and the intricacies of its parts, we cannot help but recognize our responsibilities to ourselves, to our children, and to the rest of the animate world.

My hope is that by discussing in scientific terms issues of public concern, I might begin this process of educating scientifically literate environmentalists. My goal, however, is not merely to provide answers but rather to demonstrate how to ask the correct questions. It is, after all, only by asking astute questions that we will be able to influence our elected and appointed officials along with the leaders of the corporate world.

Responses to the newspaper columns I have been writing on scientific and environmental topics for the past few years have convinced me that significant progress can be made. But an enormous amount of work remains to be done, not the least of which is to reach people before they become set in their irrational ways. When the editor of the trade journal *Pest Control* editorializes against the "Zimmermans of this world" who expect scientific documentation of pesticide safety, when creationists condemn me to hell and, for good measure, write to the president of my college in an attempt to get me fired, and when homeopaths justify dosing my ailing father with DDT, it is distressingly apparent that some people's deeply held views are at such odds with reality that little meaningful education is possible. Nonetheless, it is impossible not to be encouraged when, after publishing a column discussing the dangers of hunting faced by a diminishing New Mexican population of nearly extinct whooping cranes, I had the opportunity to meet with a second grade class and discovered that in their naive ways, each one of those students was an environmentalist. While discussing some of the problems associated with trophy hunting, global warming, and extinction through habitat destruction, they were able to recognize that wildlife has an inherent value, and they were willing to weigh that value against other economic advantages. Most remarkably, they were able to recognize that there are times when trade-offs between competing goods need to be made. If we can help those students to retain their idealism while helping to shape their critical faculties, the world will become a better place.

Education of this sort will occur only when scientists step outside their laboratories and communicate directly with the public and when the public begins to understand the beauty inherent in science and to lose its fear of the subject. In the present volume, I discuss a wide spectrum of seemingly disparate issues; chapters range from food safety to scientific fraud, from land use planning to pseudoscience. Although it is surely reasonable to ask how topics as diverse as creationism and environmentalism might be addressed in a single book, I actually see the two to be quite closely linked. For example, if someone believes that 4,500 years ago the continents were united into a single land mass upon which humans and dinosaurs lived in harmony until disrupted by a virtually instantaneous worldwide flood of such magnitude that every bit of land was under water, it seems extremely unlikely that this person is going to be particularly upset by the prospect of a 1–4° C increase in worldwide temperature occurring over a fifty year span. Pseudoscientific beliefs can thus significantly color our perception of the natural world. Similarly, an unrealistically naive view of technology coupled with a misguided reliance on its potential to mitigate human error will have equally damaging consequences. The strong, if subtle, thread tying all of these assorted subjects together is the skepticism inherent in the scientific method.

This book, then, is meant to be a primer. It is designed to introduce the nontechnical reader to the nature of science in a way that will facilitate the blending of science and environmentalism and allow readers to examine and understand the very tangible connections between our technological plateau and the contours of our planet. It is designed to provoke readers to ask increasingly sophisticated questions about how the world, locally and globally, is being used and tragically abused.

What I have tried to do in the chapters that follow is to make those connections and those questions as explicit as possible. I've also adopted and made use of a distinction that is so important in my field: the difference between natural history and ecology. Natural history is largely an observational endeavor, whereas ecology is largely an experimental one. Meaningful experiments in ecology can only occur after extensive observation. A scientist must understand an enormous amount about her study system before she is ready to begin to ask challenging questions.

In much of *Science, Nonscience, and Nonsense,* I help the reader to observe patterns, in essence, undertaking a natural history of environmentalism. Occasionally it is possible to go a step further and ask a critical question or two. This book is not meant to be either inclusive or exhaustive. Rather, by using somewhat eclectic, but nonetheless diverting, examples, I hope to educate readers about the ways we use and abuse science and the ways we reach public scientific decisions. I would like us all to have a voice in improving the quality of those decisions.

Chapter 2 draws on the premises of a number of popular pseudoscientific beliefs to document the basic philosophical underpinnings of the scientific method. Only when we are able to understand and appreciate the methodological differences between science and pseudoscience will we be able to assess to any meaningful degree the claims made by either. Chapter 3 takes the next step and addresses how we know what we believe we know. Unfortunately for many put off by the discipline of mathematics, the twin languages of science are statistics and probability. When a scientist conducts an experiment, she has to know how to interpret her results. She has to know when the pattern she discerns is due to chance or, instead, is due to a very real interaction between important variables. By using a range of examples including the statistical consequences of random drug testing, the odds of winning the state lottery, and the efficacy of treating coronary care patients with intercessory prayer, Chapter 3 is designed to familiarize readers with enough statistical background to make them more skeptical citizens.

As Chapter 4 demonstrates, a large array of political forces shapes the scientific agenda of the country. While governmental officials acting as stewards of taxpayer dollars surely have an obligation to be sure that the funds entrusted to their care are well spent, all parties need to recognize that there are limits beyond which the scientific process itself becomes perverted. Chapter 5 uses the information that has preceded it to demonstrate that many of our most severe environmental problems are not spatially or temporally localized. Rather, actions taken at one time, in one location, often have devastating effects elsewhere at a later time. Recognition of this principle is critical because temporal and spatial displacement makes it exceedingly difficult for causes to be linked to effects.

Chapters 6 to 9 focus, respectively, on endangered species, the impact of chemicals in our environment, food safety, and land use. In each case, specific examples are viewed through the lens of critical thinking in the hope that they can help determine which environmental policies make sense within a rational, scientific framework. Too often, political consideration seems to overshadow and even obfuscate scientific data. This is not to say that scientists have sole access to truth or that scientists should be the sole, or even the primary, decision makers when it comes to public policy. Rather, the risks are great when we ignore what science can teach us.

The final two chapters present a warning and a hope for the future. The warning, delivered in Chapter 10, makes it clear that we would be extremely unwise to rely on technological advances to fix our host of environmental problems. Instead, we have to search for alternative solutions and different ways of thinking about our problems. Some of these solutions are outlined in Chapter 11. As with the rest of the book, this chapter is not designed to serve as a blueprint for a better world but as a beginning toward improvements in environmental literacy.

The words of Louis Pasteur, from a speech delivered at the jubilee celebration of the Pasteur Institute in 1892, eloquently summarize my hopes for science. "[I hold] the invincible belief that Science and Peace will triumph over Ignorance and War, that nations will unite, not to destroy but to build, and that the future will belong to those who have done most for suffering humanity."

Acknowledgments

Without the wonderful editing of Abby Frucht, *Science, Nonscience, and Nonsense* would not exist.

The superb reference staff at Oberlin College was invaluable. Jeff Schloss, Steve Wojtal, and David Egloff commented on various portions of the book or provided needed information. I am grateful for their help.

Andrea Gullickson, Wendy Kozol, Estella Lauter, Ed Linenthal, Lynn Powell, Ron Rindo, Jeff Schloss, Dan Stinebring, and Steve Wojtal have been supportive in ways unimaginable. I am proud to share some of the good times with friends like these who have helped me through some of the hard times.

I also want to thank the numerous newspaper editors who have worked with me over the years and who saw enough value in my writing to provide me with a public forum. Some material in chapters 2, 9, and 11 originally appeared in newspaper columns coauthored with J. D. Zimmerman, Peter Riggs and Joan Hartmann, respectively. I thank each for fine ideas and the pleasure of collaboration.

Science, Nonscience, and Nonsense

Chapter One

～

A Personal
Reflection

Much has happened since 22 April 1970, the first Earth Day. That day of festivities solidified and energized the nascent, grass-roots environmental movement, demonstrating that people from all walks of life and all parts of the country cared deeply about the environment. And it showed that environmentalists made up a political force that could not easily be entirely ignored. Partly in response to Earth Day, we saw the creation of the Environmental Protection Agency and the Council on Environmental Quality in 1970, federal clean air and clean water acts, and the Endangered Species Act in 1973.

Not all that has happened to our planet since then, however, has been so good. Since that first Earth Day, there are more of us, generating more trash and abusing more land, than there used to be. In 1970, when we were worrying about the consequences of overpopulation, there were "only" 3.72 billion of us. Today, approximately 5.6 billion human beings are crowded together on our planet—an increase of more than 200,000 people per day. In 1970, in the United States we generated 100 million tons of solid waste yearly. By 1990, despite our environmental awareness, that figure had risen to 158 million tons. Our use of beverage cans alone has grown astronomically, from 11.5 billion in 1963 to 70 billion in 1985. And our trash problems have not been limited to earth. Orbiting our planet in 1970 were approximately two thousand pieces of debris large enough to be cataloged. By 1990 that number had grown to about seven thousand, with an uncountable number of additional, smaller, but still extremely dangerous, pieces. (These objects are defunct satellites, pieces of rockets originally used to launch the satellites, and debris created when satellites collide. Many of the objects hurtling out of control are the size of a filing cabinet. Every time a collision takes place, smaller pieces of junk are strewn about, and because each piece is traveling at about 18,000 miles per hour, ten times faster than a rifle bullet, it can cause serious damage if it strikes a spaceship or an astronaut taking a stroll.) During the decade following the original Earth Day, we turned 13 million acres of our nation's farmland into urban areas. Although specific figures are not available for the second decade, this appalling trend has continued. We continue to plunder our national forests as well. In 1970, approximately 11.5 billion board feet of timber was harvested from those forests yearly; in 1986 the figure stood at 12.6 billion board feet.

Although there are more human beings now than there were on Earth Day in 1970, we are sharing our planet with fewer individuals of other species. In the United States alone, 539 species are currently listed as endangered or threatened. In 1970 "only" 92 species were listed, one of which was the dusky seaside sparrow, a small southeastern bird. That species, having become extinct a few years ago, is no longer on the endangered list. In 1970 we coexisted with an estimated 4.5 million African elephants; today there are barely 500,000 remaining.

By Earth Day 1970 we had already been awakened to the dangers of indiscriminate pesticide use by Rachel Carson's *Silent Spring*. At that time, human fat averaged eight parts per million of DDT; in 1983 that value had dropped to two parts per million. But rather than extrapolating the lesson of DDT to other pesticides and weaning ourselves from them, we merely substituted one poison for another and became ever more dependent. In 1970 global pesticide sales were $5 billion; projected sales this year exceed $50 billion.

Perhaps the most distressing statistic is that in 1970 the federal government devoted 1.5 cents of every tax dollar spent to natural resources and the environment. In 1990, even with a self-proclaimed environmentalist in the White House and with numerous public opinion polls showing Americans to be more aware of environmental problems and more willing than ever to spend the requisite amount of money to correct those problems, the federal government under the Bush administration was still earmarking only 1.5 cents of every tax dollar to natural resources and the environment.

We've come a long way since that environmental awakening in 1970. We've seen what is probably the beginning of global warming, the creation of ozone holes over both the Northern and the Southern hemisphere coupled with an escalating rate of skin cancer, and chemical and nuclear accidents in Bhopal and Chernobyl which dwarf what most people ever believed possible. Yet, although we still live in a terribly consumption-oriented society, we have also seen the coming of the green consumer. Given a choice, many of us will now opt for products that are easy on the environment.

In 1970 I was a student in a suburban New York high school. Caring little about science, I knew virtually nothing about the environment and did not participate in that first Earth Day at all. I've come a long way since then, too. One of the things I've learned is just how important a good teacher is. Quite serendipitously, during the summer of 1971, I found myself enrolled in a science course at Hampshire College sponsored by the National Science Foundation. The course topic was the microbial ecology of compost piles. I was one of twelve or so high school students and recent graduates asking what happened to the pathogenic bacteria in sewage sludge when it was composted. Amazingly enough, before that summer, like virtually everyone else in that course, I had never heard of a compost pile, had no idea what a pathogenic bacterium was, and had never conducted a real scientific experiment. But I had an astoundingly good teacher, a microbiologist named Lynn Miller, who refused to accept our ignorance and introduced us to the beauty and elegance of the scientific method. He also taught us about skepticism and critical thinking. That summer he badgered us by asking not what we knew but how we knew it. He taught us that a good question was every bit as important as a good answer. He also taught us that there were no shortcuts to true knowledge. In just four short weeks, Lynn's enthusiasm and his lessons persuaded me to become a scientist.

That dream stayed with me through my undergraduate and graduate years; ultimately I earned a Ph.D. in ecology, studying the reproductive biology of *Polemonium foliosissimum*, a Rocky Mountain herb pollinated by bumblebees. Soon thereafter I was lucky to find a wonderful job teaching biology and conducting research at Oberlin College.

Throughout all of this time, although I was a professional ecologist, I was not really an environmentalist; my sensibilities were in the right place, but my actions were nonexistent. My research continued in Colorado on the foraging strategies of bumblebees and on the reproductive patterns of the plants they visited. Although I heavily involved my students in my research, I was the typical ivory-towered professor—my work did not intersect with the real world. (I say this not as a criticism but as a fact. My research findings, in fact, could be said to have some applied aspects, two steps down the road, but I was much more interested in them for their own pure sake.)

My world began to change slowly on 19 March 1981. It was on that day that Governor Frank White of Arkansas signed Act 590 of 1981 into law. The heading of that bill reads as follows:

> An act to require balanced treatment of creation-science and evolution-science in public schools; to protect academic freedom by providing student choice; to ensure freedom of religious exercise; to guarantee freedom of belief and speech; to prevent establishment of religion; to prohibit religious instruction concerning origins; to bar discrimination on the basis of creationist or evolutionist belief; to provide definitions and clarifications; to declare the legislative purpose and legislative findings of fact; to provide for severability of provisions; to provide for repeal of contrary laws; and to set forth an effective date.

Like most evolutionary biologists, I had been aware of a number of small creationist groups and their very odd writings for a number of years, but I looked on them more as an anachronism than anything to be taken seriously. A state law mandating the teaching of creationism, a law that was much more in tune with the antievolution crusades of the early part of the century than the swiftly moving biological world of the latter part of the century, however, was something to take very seriously indeed. I started reading a bit more widely on the topic and began to develop a small creationist library. After reviewing some material for the American Civil Liberties Union's court challenge to the Arkansas law, I was hooked and developed a college-level science course for nonmajors using the creation-evolution controversy as a means of discussing the nature of science and the interaction between science and society.

My response to the creationists was so visceral because their polemics so directly contradicted everything I held to be intellectually dear. Not only did they dismiss the scientific method, they blatantly misrepresented the truth. The fabrications that the creationists have been perpetuating about *Archaeopteryx* is a perfect case in point. The August 1988 *Bible-Science Newsletter,* a publication of the Bible-Science Association, one of the leading creationist organizations, contained a long interview with creationist Ian Taylor. In a section subtitled "Queer Bird," the implication is clearly that the publication of Charles Darwin's *On the Origin of Species* led to rampant speculation about what an intermediary between a bird and a reptile might look like and that this speculation led, almost immediately, to the concoction of fossils of *Archaeopteryx.* Conveniently ignored in this scenario was the fact that the first *Archaeopteryx* fossil was found in 1855, four years prior to the publication of the *Origin.*

This misrepresentation is insignificant compared with what is stated in the final paragraph of the "Queer Bird" section. This paragraph is set up by reporting that astronomer Sir Fred Hoyle believes "that the *Archaeopteryx* fossils are really fossils of a small dinosaur called a *Compsognathus,* with feather impressions stuck on by a forger." The final paragraph then goes on to read, in full:

> Since Sir Fred Hoyle published his finding of the *Archaeopteryx* fossils, the British Museum has closed off all further investigations of its specimen—an instructive example of the open, scientific attitude. Interestingly, neither Darwin nor Huxley nor any of the other early evolutionists accepted *Archaeopteryx* as a genuine transitional fossil. Maybe they knew something modern evolutionists don't know.

The point is, however, that the British Museum did no such thing. A large-scale study, using a variety of techniques, was conducted by British Museum staff *after* Hoyle and his co-workers raised their charges. The results were published in *Science* in 1986, more than two years before the interview was published in the *B-S Newsletter.* Now, I would have no problem whatsoever had Taylor decided to criticize the study or to disagree with its conclusions or methodology. Such disagreements are, after all, what science is all about. But for Taylor to say that no such studies were ever conducted and that the British Museum refused to permit such studies is an outright lie. It makes for a better story to say what Taylor did, but it just isn't the truth. Yet many creationists continue to spread this story about *Archaeopteryx.*

In addition to this sort of outright fabrication, the creationists were preaching, so unlike what I had learned from Lynn years ago, that there were easy answers and, in fact, that there were questions that should not be asked. As an educator I felt that my job was to encourage students to think thoughts they never previously imagined, to think critically and skeptically. My job, in other words, was to open students' minds, not to hammer them shut. The specter of creationism reentering the public schools was a serious threat to the very core of science education, a threat that shook me up so much that I began writing and speaking publicly about it. By leaving my classroom and laboratory occasionally for a more public forum, I began to encounter other forms of pseudoscience as well. Like creationism, most of these pseudosciences, disciplines such as homeopathy, graphology, and astrology, also promised easy solutions to complex problems, and like creationism, most of these pseudosciences were antithetical to the scientific method. In response, I began to broaden the range of my writing and speaking and began to think seriously about the ways members of a democratic society are able to put science to productive uses. My background as an ecologist and my longstanding interest in environmentalism naturally led me to focus most heavily on the interface between public science and the environment. Although the links between creationism, other forms of pseudoscience, and environmentalism might seem tenuous to some, in the case of my personal development they were quite strong. In fact, if you accept that sound environmental decision making should be built on a solid scientific foundation and that pseudoscience attacks the scientific method, the links are quite evident.

As an educator I have to believe that education can make a difference. I also can't help but believe that the more people know about the world around them, the more likely they are to make rational policy. And I know that it is possible to present technical information in a form accessible enough that nonspecialists can grasp its meaning. I know from my personal experiences as a student the dramatic effect that an exciting teacher can have. And I know from reports coming out of Taiwan a few years back that it is possible for all of this to come together at the community level as well. Taipower, Taiwan's state-operated power corporation, was convinced of Taiwan's need for a new nuclear power plant. With a majority of citizens opposed to the idea, however, Taipower thought that it had best develop a base of public support before it moved forward with the project. The popular view among those in the nuclear industry in the United States has long been that only an uninformed public is opposed to nuclear power. Counteract the high level of ignorance, the thought goes, and support for nuclear power will swiftly rise.

Taipower thus undertook a massive educational campaign. In just over four months, it spent more than $460,000 to educate the public. It staged more than 150 lectures, debates, and discussion groups in schools and cultural centers and produced a series of television programs promoting the virtues of the proposed plant. If success is to be measured by a shift in public opinion, Taipower's education plan was a success; as knowledge replaced ignorance, more and more citizens decided that they were against nuclear power. And it should be pointed out that this shift against nuclear power occurred in the face of significant bribes offered by the utility. It offered to compensate individuals living near the proposed plant by more than $6 million a year during the ten-year construction phase and over $4.6 million annually during the operating period. Still, the Taiwanese people, having been given the opportunity to evaluate technical information about nuclear power, reached the rational decision that it was in neither their economic nor their environmental interest to proceed.

Seven years ago, while I was still living in Oberlin, Ohio, I wrote the following essay. I believe now, as I did then, that it demonstrates the distance many people have placed between themselves and virtually any form of "nature." Rather than attempting either to improve on it or to paraphrase it, I present it as it appeared in *The Gamut: A Journal of Ideas and Information* published by Cleveland State University.

A Boy and a Dog and a Lawn: A Perversion of Values

A boy and his dog. Although a mite sexist, that phrase conjures up an image that is as all-American as motherhood, the flag, and any apple pie. Ten years ago my wife and I (although she wasn't my wife then) started down the road toward fulfilling that image. We adopted an abandoned dog (a mostly beagle) at a truck stop in Missouri. It took us another seven years (although, to be honest, we weren't actually trying during that period) to finally add the boy. Everything was great. Sure, Alioth (the dog) and Alex (the boy) virtually ignored each other for three years, but all parts of the image are now in place.

And then this evening we had to face the real problem. Yes, like most serious problems, there had been warning signs, and, yes, we ignored them. This evening the four of us were taking an after-dinner stroll around Oberlin, Ohio, the small college town in which we live. Alex, pretending that he was Alioth, walked slightly ahead of us, the leash attached to the waistband of his pants; Alioth ran free.

And then it happened. A gentleman in a police car called me over, introduced himself as the town's dog warden, and asked if I was aware of the new pet law in town. I pleaded ignorance. Predictably, he noted that all pets had to be leashed if they were off their own property and commented that although we were using a leash it seemed to have the wrong item attached. After a chuckle, I asked about all of the locations where people had, for years, taken their dogs to exercise and play. I asked about the reservoir, the town green, the arboretum, the school yard. Nope, even those traditional spots were covered; dogs had to be leashed.

When I asked for his suggestion about giving Alioth some exercise, he suggested running alongside her, leash in hand. Alioth loves to run in circles so, even if I were in better shape than I am, I didn't think I could handle that.

Then the dog warden came to the heart of the law. He claimed that he would have to ticket anyone whose pet defecated or urinated anywhere in town unless the act occurred on the owner's property. I just laughed, but he said that he was serious.

Now a dog's bladder has to be an interesting organ; it simply never goes dry. When she wants to mark something, Alioth never has trouble, regardless of how many things she's already marked. And marking is instinctive—it's a dog's way of demonstrating ownership. There's just no way to keep a dog from being territorial; at least no way that would win the approval of the ASPCA.

The warden was adamant. If your dog urinates anywhere other than in its own yard, you get a ticket. A second "offense," a second ticket. Ticket prices increase through the third urination, while the fourth wins you a trip to court. It was very clear that our family strolls were going to become expensive.

I understand the rationale behind leash laws, although I may not agree with them. I even understand "pooper scooper" laws, although I find them demeaning to the participants on both ends of the leash. But a law making it illegal for a dog to urinate is absurd. (Sure, all of us have had the unpleasant experience of struggling to scrape the bottom of our shoes clean after a misstep. But how many people have ever had to wipe dog urine off their shoes? Has anyone ever stepped ankle deep in a puddle?) I wonder why the members of the Oberlin city council didn't have the nerve to be honest about their intentions. Why didn't they directly enact a law making it illegal to walk a dog in town, since that is the effect of their law?

Numerous studies have shown that dogs repay their owners in many ways. Health, both physical and mental, is improved by pet ownership. Dog owners get more exercise than do non-owners. And there's that boy. A pet teaches kids about responsibility. (I'm sure you've seen the ad suggesting that virtually all of our astronauts started out as paperboys. Well, virtually all of them had dogs, as well.) But if you can't take the dog off your lawn, how can you romp, how can you play, how can you even own one? And what about those people who live in apartments and don't have lawns?

O.K., by now you're thinking that the situation in Oberlin, although extreme, is unique and that this really couldn't affect you. That may not be as true as you think. Remember, I said that the early warning signs had been present for some time. Within the last year my wife and I have been scolded in Ohio, Colorado, Kansas, and New York for allowing Alioth to urinate on someone's lawn. Please note that I said urinate, not defecate. Oddly enough, at least one of the people complaining had a dog of her own. In New York we were even vociferously chastised for allowing Alioth just to walk on one suburban lawn. And every piece of public land that we encountered there—parks, school yards, beaches, etc.—had signs saying: "No Dogs."

We live in a bizarre society when people apply massive amounts of chemicals to their lawns and worry about a drop of dog urine. Increasingly, I'm seeing little signs saying: "Chemical Treatment—Stay Off Grass Until Dry." Something has to be wrong when people are not outraged that our federal government refuses to acknowledge the links between industrial pollution, acid rain, and the devastation of plants and animals, or that no one complains about the subsidies we pay to tobacco farmers, but they are incensed when a dog leaves a drop of a non-toxic liquid that is 95% water on their lawn. I'm afraid that many people must be out of touch with reality. These laws and people's actions reflect their feelings about nature. Such feelings cannot be healthy either for them or for society.

Perhaps people are just more similar to dogs than they would like to believe. Perhaps this antipathy to urine is the way property owners can exercise their own territorial rights.

A boy and his dog. The latter, by her very nature, is turning the former into a criminal every time they go out for a walk. What am I to do?

At least the dog warden only works two hours a day in Oberlin, and our friends, whose two-and-a-half-year-old son liked to urinate on the grass, just moved out of town.

I believe that we can change attitudes like those described above. As the Taiwan nuclear power example demonstrates, education and knowledge can make a huge difference. We have to educate people fully about the dire consequences of certain attitudes, and we have to educate people about the real consequences associated with changing them. One reason many people, including politicians, are nervous about adopting an environmental ethic is their belief that to do so would entail great sacrifice. That sound environmental policy comes with a huge price tag, for example, has become virtually accepted as gospel. Accordingly, environmentalists who strongly advocate pollution abatement projects, a halt to the destruction of tropical forests, or a host of other large-scale activities are often characterized as economically ignorant idealists. Two studies published in 1989, the first two of their type ever undertaken, have shown that the gospel is wrong and that the environmentalists are not so ignorant after all.

The first analysis, published in *Nature* and conducted by Charles M. Peters, Alwyn H. Gentry, and Robert O. Mendelsohn, botanists at the New York Botanical Garden, the Missouri Botanical Garden, and Yale University respectively, sought to identify the economic value of a tract of Amazonian forest. The scientists began by exhaustively sampling every tree in one hectare (approximately 2.5 acres) of forest in Amazonian Peru. The diversity of trees encountered was staggering— 842 trees greater than four inches around were found, representing 275 different species and 50 families. Almost 42 percent of the trees yielded products marketable in the region. The smaller vegetation consisted of many plants of medicinal value. The marketable products present on that small piece of land were of different sorts: lumber, fruits, vegetables, herbs, and rubber.

What would be the economic gain of various land use strategies?

The long-term economic consequences of three strategies were examined: clear-cutting for commercial timber; annual harvesting of the renewable resources (fruits, vegetables, rubber, and so on); and farming. The farming option was immediately ruled out on economic grounds because it is well known that as lush as these tropical forests are, when cleared, the land they were growing on cannot remain in agricultural service for any appreciable length of time because of incredibly thin soil profiles. (Owing to the very high rates of decomposition in the tropics, most of the nutrients in tropical forest ecosystems are tied up in the plants rather than in dead and decaying matter on the ground.) When all economic costs and benefits of the other two options were taken into consideration, the botanists came to the surprising conclusion that the land would yield about thirteen times as much income if the renewable products of the trees rather than the trees themselves were harvested. When the medicinal value of the smaller plants was factored in, the differential was even greater. And the botanists' calculations purposefully ignored the nonfinancial benefits of keeping tropical forests in place, benefits ranging from mitigation of the effects of global warming to the scientific and aesthetic importance of these ecosystems. When environmentalists work to halt the current and precipitous destruction of tropical forests, they now have data proving that a change in policy, a change *toward* preservation, is in everybody's economic self-interest.

The second landmark study, published in *Ambio* in 1989 and conducted by three Washington economists, focused on the costs and benefits of pollution abatement controls. As with the Amazonian analysis, results showed the gospel to be completely incorrect: the study did "not lend support to the widely held belief that environmental programs generally hurt the economy by crippling industries and increasing unemployment." The results, in fact, showed quite the opposite.

The pollution abatement control industry itself has become a big business that produces major capital equipment while providing large numbers of jobs. The three economists showed, for example, that the $8.5 billion spent by industry in 1985 to control air, water, and solid waste production, when coupled with contributions made by government, translated into pollution abatement industry sales of $19 billion, profits of $2.6 billion, and 167,000 new jobs. Although it cannot be disputed that pollution abatement expenditures will cost the polluting industries some profits and cause the loss of some jobs, the analysis makes it clear that dollars spent on such items represent a net economic gain to society. And like the study of the Amazonian forest, this analysis is purely economic, factoring in only the monetary benefits of pollution abatement. It would, after all, be difficult to place a monetary value on breathable air and potable water.

The two studies are important because they challenge the basic economic assumption that has for years served as the fundamental criticism of the environmental movement. Now that that assumption has been found seriously wanting, and now that the current economic analyses agree with what ethically, morally, and environmentally we have always known to be the case, perhaps we can make significant environmental progress.

Although he was speaking explicitly about the problems associated with the extinction of species, E. O. Wilson, in his latest book, *The Diversity of Life,* very eloquently explained just how large the stakes are: "If enough species are extinguished, will the ecosystems collapse, and will the extinction of most other species follow soon afterward? The only answer anyone can give is: possibly. By the time we find out, however, it might be too late. One planet, one experiment."

Chapter Two

༄

Scientific
Illiteracy

From "Creation
Science" to
Graphology

On 22 February 1923 Oklahoma state representative J. L. Montgomery introduced an amendment to a bill designed to provide free textbooks to public school students throughout his state. His amendment prohibited the state from distributing any text that favored the "Darwin Theory of Creation" over the "Bible Account of Creation." In defense of his amendment, which passed overwhelmingly, Montgomery pronounced, "I'm neither a lawyer or a preacher, but a two-horsed layman and I'm against this theory called science!"

A 1991 congressional debate shows that, unfortunately, such extreme views are not limited to historical anecdotes. In discussion of a relatively small item in NASA's budget greatly upgrading its program in search of extraterrestrial life, Representative Ronald Machtley (R-R.I.) advised his colleagues that "money ought not be spent on curiosity" (quoted in *Science,* 20 July 1991, 249). This extremely peculiar view (what, after all, are scientists, not to mention other human beings, supposed to be if not curious?) apparently was compelling: the House voted to ax the entire project.

These two examples demonstrate a serious misunderstanding about the very nature of science. Although he may have been unable to articulate it, what Representative Machtley probably meant was that we should be spending our funds on applied rather than pure research. The former, by definition, typically manifests immediate benefits to society, whereas the latter often appears frivolous. As I discuss more fully in Chapter 4, this rather common perception is terribly shortsighted. Only on a solid base of pure research, that is, research with no immediate applications, does applied work ever become possible. Who would have predicted sixty years ago that the purely theoretical research by physicists studying the effects of magnetic forces on the behavior of nuclei might evolve into the technology of magnetic resonance imaging (MRI), which has so dramatically altered modern diagnostic medicine?

There is another deeper, and more important, level of misunderstanding about the nature of science reflected in Machtley's comments. He, like so many others, seems to believe that science is a collection of facts rather than an ongoing investigative process that permits us to gain insight and understanding into the way the world works. Interestingly, Montgomery's remarks are probably more indicative of a meaningful understanding of what science is really all about. Although he might not have fully thought through the philosophical issues, a modern investigative definition of science incorporating both inductive and deductive logic as well as a heavy dose of skepticism is something one might oppose, especially if one is a fundamentalist secure in the belief that scientific truths have been presented to humans by an omnipotent and omniscient god.

It can hardly be disputed that the United States is in the midst of a crisis in science education when international tests across a wide range of scientific disciplines repeatedly place American students behind those of virtually every other developed country. The problem is less that most Americans share no solid grasp of a body of scientific "knowledge" (although many surely do not) than that they have a complete misunderstanding of the nature, processes, and purposes of science. Americans lack the critical capacity to distinguish real science from pseudoscience. A few recent findings demonstrate the extent of this illiteracy and document that even "well-educated" people are not immune:

— Jane Brody, in the *New York Times* on 9 June 1988, wrote, "A Louis Harris poll conducted for the FDA [the U.S. Food and Drug Administration] in 1986 among more than 1,500 adults revealed that 26 percent of Americans used one or more questionable methods of health care to combat one of several medical problems especially prone to quackery."

— Raymond Eve and Francis Harrold of the University of Texas at Arlington sampled student opinions at their institution. They reported that only 29 percent of the students do not believe in reincarnation, only 36 percent reject the idea that communication with the dead is possible, and only 22 percent are confident that the future cannot be predicted by psychic power.

— In a study of the managing editors of the nation's daily newspapers, I found that only 57 percent of the 534 respondents disagreed strongly with the statement "Every word in the Bible is true" and only 51 percent disagreed strongly that "dinosaurs and humans lived contemporaneously."

What we have is a public largely anxious to jump to supernatural and so-called alternative conclusions. The scientific community's commitment to the scientific method in which cause and effect have to be tightly linked has been misinterpreted by much of the public to mean that scientists narrow-mindedly refuse to consider views diverging from the accepted norm. Such a public is unable to make scientific decisions based on evidence rather than emotion. And in a democratic society such as ours in which a huge proportion of the funding for scientific research comes, at least indirectly, through the actions of Congress, the public has a large say in setting the country's scientific agenda. It is thus imperative that basic misperceptions about the nature of science be overcome.

One of the best ways to articulate those misperceptions is to compare mainstream science with clearly established cases of pseudoscience. The most successful example of pseudoscience has clearly been the "creation science" movement of the past twenty years. (I should note that "creation science" has been put in quotation marks because, as I will demonstrate, the creationists have attempted to appropriate the terminology of science while undermining its very methodology. Simply put, there is nothing resembling scientific investigation in "creation science.") By successful, I mean that "creation scientists" have been remarkably adept at shaping public school science curricula nationwide and fashioning public opinion in a light very favorable to themselves. Polls have repeatedly shown that upwards of three-quarters of those surveyed are in favor of creationism being taught in the nation's public schools and 10–16 percent prefer that *only* the creation model be taught. The creationists have been halted from completely overturning and redefining mainstream science instruction only by a Supreme Court ruling in June 1988.

Although they may not be particularly sophisticated scientifically, the leaders of the modern-day creationist movement are extremely politically astute. The slogan they have adopted, freedom of speech, has caught the attention of the general populace and in the mid-1980s was endorsed by President Ronald Reagan. The media continue to run feature articles and op-ed pieces advocating the teaching of "creation science" in the public schools, arguing that not to do so would be a form of censorship. Proponents of such a viewpoint typically cloak themselves in the U.S. Constitution and behind such notables as Thomas Jefferson, John Milton, Oliver Wendell Holmes, Louis Brandeis, George Santayana, and John Stuart Mill.

In fact, to equate the teaching of "creation science" in public schools with freedom of speech is like saying that palm readers should have their say, too, for the creationists' arguments remain simplistic and misleading. They usually advocate two points. First, they say champions of academic freedom and truth should solidly support state laws or local school board mandates that dictate equal time for "creation science" because such rules provide for a free and complete discussion of all pertinent issues in public schools. Second, they claim that without such rulings, revision or rejection of the theory of evolution would be impossible because of what they call the close-mindedness of the scientific community.

Let's start with the most basic point: the definition of "creation science." For the general public, "creation science" has, rather recently, been sanitized and redefined by its proponents as the body of scientific evidence for, not the biblical account of, creation, along with inferences drawn from that evidence. The in-house writings of the creationists, however, tell a very different story. Henry Morris, head of the Institute for Creation Research (ICR), the world's leading creationist think tank, disagrees with this redefinition. In his book *Studies in the Bible and Science,* he states: "If man wishes to know anything about Creation, . . . his sole source of true information is that of divine revelation. God was there when it happened. We were not there. . . . Therefore, we are completely limited to what God has seen fit to tell us, and this information is His written Word. This [the Bible] is our textbook on the science of Creation!" Morris goes on in *The Remarkable Birth of Planet Earth* to state: "The only way we can determine the true age of the Earth is for God to tell us what it is. And since He has told us, very plainly, in the Holy Scriptures that it is several thousand years in age, and no more, that ought to settle all basic questions of terrestrial chronology."

Or consider these comments from Ken Ham, a leading creationist spokesperson and staff member of the ICR, published in the July 1992 issue of ICR's *Acts and Facts:* "Because the Word of God is the only absolute we have, all teaching must be checked against this absolute standard. We here at ICR have always tried to be true to Scripture. We put God's Word first; we build our science on God's Word, not on our understanding of God's Word based on science."

The scientific views of the "creation scientists" can perhaps be best summarized by the bumper sticker that says: "God said it, I believe it, That settles it." When the biblical references are removed from "creation science," not much remains. Three examples should make this point:

— Before an Arkansas law demanding equal treatment for evolution and "creation science" was passed in 1981 (and declared unconstitutional in 1982), the Little Rock Board of Education formed a committee to draw up a curriculum guide dealing with creationism. The committee was frustrated in its endeavor because while it repeatedly encountered biblical references in creationist literature, it could find no scientific ones.

— A draft curriculum guide including creationism was prepared for the Columbus, Ohio, public schools in August 1981. Faced with a lack of real science in "creation science," the guide was forced to cite articles published in the *National Enquirer.*

— W. Scott Morrow, a creationist testifying in favor of the Arkansas equal-treatment law, said, under oath, that "creation science" was nothing more than "an accumulation of asserted inconsistencies or insufficiencies in the evolutionary model."

In short, "creation science" cannot be divorced from the Bible without completely evaporating. There is no science in "creation science."

What about the claim of academic freedom? Who, the creationists ask, could be against the schools continuing as a free marketplace of ideas?

The answers to these questions are really quite simple. First, equal-time laws do not allow a multitude of ideas the opportunity to compete with one another openly. Rather, they dictate only that creation according to fundamentalist Christian ideology be discussed. Second, academic freedom does not require that all ideas be given equal weight. Most of us would agree that some thoughts are more important than others or that some ideas are well founded and others are not. Do the tenets of academic freedom mean that even incorrect ideas should be presented? Of course not. For example, Copernicus and Galileo demonstrated that the earth revolves around the sun and that the sun is the center of our solar system. The geocentric theory of the universe (that the earth is the center of the universe) has not been taught for centuries. Clearly, public school teachers do not and should not present both sides of this debate, except when referring to the history of science.

But perhaps the issue isn't quite that clear. A number of vocal creationists insist that geocentrism is a fact and demand that it be taught in public school science lessons. At the National Bible Science Conference in 1984, several speakers quoted Scriptures to demonstrate that the sun does indeed revolve around the earth. Not one of the "creation scientists" present objected to this ridiculous view of astronomy.

Perhaps even more absurd is the situation that arose in February 1987, when Evan Mecham was still governor of Arizona. Jim Cooper, Mecham's chief aide for education matters, testified before a legislative committee that public school teachers should not impose their belief that the earth is round on students who have been brought up to believe that it is flat.

Equal-treatment laws could well require that geocentrism and perhaps flat-earthism, both of which long ago lost out in the free marketplace of ideas, be taught as alternative views in science classrooms. Likewise, the biblical theory of creation lost to the scientific theory of evolution in the nineteenth century. The creationist cry for academic freedom serves only to artificially preserve outdated, stale, fanatical concepts. Vital, meaningful scientific theories continue to stand on their own merit.

The last generation of creationists was also in court protecting their biblical view of "science." At their behest in the 1920s, numerous state laws prohibited the teaching of evolution. (The last of these statutes was ruled unconstitutional by the U.S. Supreme Court in 1968.) The creationists' cry for an open discussion of all ideas rings hollow when it is seen that they have historically favored the banning of evolutionary teaching but now attempt to mandate the teaching of their outdated concepts.

What about the claim that without the challenge of "scientific creationism" the evolutionary sciences could never be proved incorrect? Let's look at the evidence.

First, a massive study by Eugenie Scott and Henry Cole published in 1985 in the *Quarterly Review of Biology* demonstrated that creationists have provided no meaningful challenge to the evolutionary sciences. Scott and Cole surveyed the editors of sixty-eight science journals to determine what proportion of total manuscripts submitted from 1981 to 1983 were of a creationist nature. What made Scott and Cole's study so powerful was that, instead of merely looking at what scientific journals published over the years, they asked editors about manuscripts that had been submitted by authors. The results were striking. Of the 135,000 only 18 (0.01%) could be categorized as falling in the creationist camp, and those appeared to be written by laypeople, not professional scientists.

In the face of empirical evidence of this sort, creationists can no longer voice their longstanding claim that editorial bias is keeping their work from being published. The only reasonable interpretation of the Scott and Cole study is that "creation scientists" are not actively doing science. They perform no experiments to test their hypotheses, they write no papers detailing their results, and they share no findings with the scientific community. The fact that creationists make no attempt to share their ideas should lay to rest their claim that the scientific community is close-minded and unwilling to listen to alternative views.

From the perspective of the nature of science, an examination of the creationists' second claim is even more instructive. Remember, they assert that without the challenge from "scientific creationism" the evolutionary sciences would be stagnant. But by their very nature, all scientific theories are constantly being reevaluated and discussed in the scientific literature. Scientists are trained to be skeptical and to question the fundamental premises of their disciplines. Members of all major "creation science" organizations and science professors at virtually all creationist institutions of higher education (e.g., Bryant College, Christian Heritage College, Liberty Baptist University), on the other hand, must sign yearly oaths attesting to their beliefs. As federal district court judge William R. Overton pointed out in footnote 7 of his 5 January 1982 opinion striking down the Arkansas equal-time law:

> Applicants for membership in the CRS [Creation Research Society] must subscribe to the following statement of belief: "(1) The Bible is the written Word of God, and because we believe it to be inspired thruout (sic), all of its assertions are historically and scientifically true in all of the original autographs. To the student of nature, this means that the account of origins in Genesis is a factual presentation of simple historical truths. (2) All basic types of living things, including man, were made by direct creative acts of God during creation Week as described in Genesis. Whatever biological changes have occurred since Creation have accomplished only changes within the original created kinds. (3) The great Flood described in Genesis, commonly referred to as the Noachian Deluge, was an historical event, world-wide in its extent and effect. (4) Finally, we are an organization of Christian men of science, who accept Jesus Christ as our Lord and Savior. The account of the special creation of Adam and Eve as one man and one woman, and their subsequent Fall into sin, is the basis for our belief in the necessity of a Savior for all mankind. Therefore, salvation can come only thru (sic) accepting Jesus Christ as our Savior."

Similarly, the Statement of Faith that must be signed by members of the Bible-Science Association includes a list of what members must accept: "Belief in Special Creation; Literal Bible Interpretation; Divine Design and purpose in Nature; a Young Earth; a Universal Noachian Flood; Christ as God and Man—Our Savior; Christ-Centered Scientific Research."

Oaths of these sorts are completely antithetical to the very nature of scientific investigation. Indeed, compare these oaths with the paraphrase of a line from Horace's *Epistulae,* which was adopted as a motto in the 1600s by the Royal Society of London, the oldest and one of the most respected of the world's scientific societies:

> I am not bound to swear allegiance to the word of any master.
> Where the storm carries me, I put into port and make myself at home.

Meaningful scientific investigation only occurs when the investigator has an open mind. Creationists, like so many other pseudoscientists, cannot be considered scientists because they state quite clearly up front that their opinions are fixed and are not open to change even in the face of contrary data.

Philosophers of science have spent enormous energy defining the parameters that are necessary for any undertaking to be considered scientific. Instead of entering that rather technical debate, it is worth looking at Judge Overton's ruling in the Arkansas equal-time case, since he specifically addressed exactly this point:

> Section 4(a) lacks legitimate educational value because "creation science" as defined in that section is simply not science. Several witnesses suggested definitions of science. A descriptive definition was said to be that science is what is "accepted by the scientific community" and is "what scientists do." The obvious implication of this description is that, in a free society, knowledge does not require the imprimatur of legislation in order to become science.
>
> More precisely, the essential characteristics of science are:
> (1) It is guided by natural law;
> (2) It has to be explanatory by reference to natural law;
> (3) It is testable against the empirical world;
> (4) Its conclusions are tentative, i.e., are not necessarily the final word; and
> (5) It is falsifiable.

In accepting this five part definition of science, Judge Overton drew heavily on the testimony of philosopher of science Michael Ruse and recognized the process nature of scientific investigation. The two words that are most important in Overton's characterization are "tentative" and "falsifiable." The tentativeness of scientific conclusions is what makes being a scientist and studying about science so much fun. It is also what makes being a scientist and a student of science so difficult. Because our understanding of the natural world is always open to improvement and modification, we cannot rely with certainty on scientific explanations. Our scientific explanations and our theories are only as good as the underlying data; with more refined experimentation our worldview might shift. Basic Newtonian mechanics, for example, was accepted as the "correct" mechanism to explain the movement of the planets until Einstein came up with a package that better fit the empirical data. The concept of tentativeness does not mean that we should lack confidence in science but rather that we need to understand that the scientific method allows our perception of scientific "truth" to mature as our data base matures.

Similarly, the concept of falsifiability brings real life to the process of scientific investigation. It means that every scientific idea must be framed in such a way that experiments can be designed which have the possibility of demonstrating that the idea is incorrect. A critical point to recognize is that there is a world of difference between an idea being falsifiable and actually being false. For an idea to be falsifiable, someone must be able to create an experiment which, if carried out correctly, could yield data that contradict predictions previously made. For an idea to be demonstrably false, one of those experiments must yield data that actually contradict those predictions.

Perhaps the best example of an experiment falsifying a hypothesis is the classic one performed by Louis Pasteur in 1864. The results of his experiment forced the scientific community to reject the hypothesis of spontaneous generation, the idea that life might regularly arise from inanimate constituents. Pasteur boiled a nutrient-rich solution and placed it in a flask with a long, curving neck. This apparatus, although open to the air, kept all airborne particles from landing on the surface of the broth. After a full year with no microorganisms present in the broth, he snapped off the neck of the flask. Now directly open to the air, the broth became cloudy and teeming with microorganisms within a mere twenty-four hours. This simple, elegant experiment effectively put an end to the theory of spontaneous generation.

Taken together, what all of this means is that science proceeds by disproving rather than proving hypotheses. There is, after all, no way to be certain that a better explanation for any particular phenomenon will not come along in the future. Thus all scientific explanations must be accepted only tentatively. On the other hand, a carefully crafted experiment like Pasteur's can conclusively demonstrate that a particular idea is entirely incorrect. Needless to say, this constant attempt to disprove ideas makes scientists a very skeptical bunch.

Given the oaths that people must sign to become active members in creationist organizations, and given the creationists' disdain for working with falsifiable hypotheses (not to mention their disdain for natural law—they often argue that the speed of light has been slowing significantly in recent years, that radioactive decay used to occur at a faster rate, and that death was not present on the early earth), it is impossible not to agree with Judge Overton's opinion that "creation science" "fails to follow the canons of defining scientific theory."

The fundamentalist attack on evolution arises in part because quality science instruction must focus on process rather than on facts. The best teacher is not the one who stands in front of a class and recites "facts" but rather the one who encourages students to ask questions and to think critically. This is in direct opposition to the main premises held by fundamentalists. As Henry Morris said, all of the scientific answers that we need have been given to us in the King James version of the Bible. When scientific answers are viewed tentatively and the Bible is viewed as a scientific text, the teaching of evolution does tend to undercut a belief in a literal interpretation of the Bible.

To many people, the skeptical thinking associated with quality science instruction is dangerous because they imagine that it undermines authority. A teacher, after all, cannot have all the answers if answers to the most interesting questions are not known. Related to this is the strong belief held by many people that the acceptance of scientific knowledge dictates both morality and public policy. Again, the creationists have been in the fore of the group holding this misguided view. Thus, the second reason why they want evolutionary teaching in public schools balanced by "creation science" or want it completely excluded from the curriculum is that they believe it has led to a decay of American society.

Although this last statement might sound extreme, the vehemence of the creationists on this topic is truly amazing. Historian Ronald Numbers pointed out that the Reverend T. T. Martin, a leading fundamentalist minister and antievolutionist activist, wrote in 1923 that the German soldiers who killed Belgian and French children with poisoned candy were angels compared with the teachers and textbook writers who corrupted the souls of children with false teaching and thereby sentenced them to eternal death.

Much more recently, in 1977, the Creation Science Research Center claimed that its research had proved that the teaching of evolution fostered "the moral decay of spiritual values which contributes to the destruction of mental health and . . . [the prevalence of] divorce, abortion, and rampant venereal disease." Not surprisingly, no data were presented to support such a provocative statement.

Still, you might say, 1977 was quite a while back, and times change.

Indeed they do. Since that time, venereal disease has been swept from the front pages of newspapers by the AIDS epidemic, and the creationists have kept pace. Ken Ham wrote in April 1987 that "we are not experiencing an AIDS crisis—but a morality crisis. The spread of AIDS can be stopped—by simply rejecting false evolution and trusting in the Creator, the Lord Jesus Christ, and by obeying the laws He gave to us."

Such mixing of science and morality is dangerous for public welfare. In a democratic society, students—indeed, the public—should have access to the best available scientific information. At the same time, scientists should have no more say than any other group when public policy, even public policy dealing with scientific and technological issues, is being formulated. Although scientists may have expertise in scientific "facts," and although that specific knowledge will often be extremely useful in the search for solutions to a range of human problems, this does not mean that scientists alone should be entrusted with political decisions relevant to that knowledge. Rather, they should make every attempt to communicate that knowledge to the broader enfranchised public. Similarly, political opinion should be irrelevant when dealing with the scientific principles underlying particular issues.

Consider a specific situation. Although we might disagree about the value of nuclear weapons, no one is advocating that we stop teaching chemistry and physics. The basics of both chemistry and physics are value free and thus must be taught. We can, and should, debate their applied uses.

The principles of evolutionary biology are no different. They are central to an understanding of all life around us and thus are necessary if a meaningful, solidly based environmental ethic is to be formed. Indeed, in 1973 the great geneticist Theodosious Dobzhansky wrote a paper entitled "Nothing in Biology Makes Sense except in the Light of Evolution." Yet because the principles of evolution have entered into, and been perverted within, the political arena time and time again, we are faced with the specter of a movement intending to discredit those principles.

The antievolution crusades of the 1920s are of particular interest in this light because two extreme political blocs joined forces. Biblical fundamentalists were against evolutionary teaching for the reasons outlined above. Many social progressives opposed it as well because of the evils that came from the laissez-faire Social Darwinism of nineteenth-century sociologist Herbert Spencer. How could anyone concerned with social welfare not be appalled after reading William Graham Sumner's words from his book *The Challenge of Facts and other Essays* written in 1914: "The millionaires are a product of natural selection, acting on the whole body of men to pick out those who can meet the requirement of certain work to be done. . . . They get high wages and live in luxury, but the bargain is a good one for society." Sumner went on to say that without inequality, survival of the fittest would have no meaning. Social Darwinism, we know now, was a perversion of biological principles. Although the biology was sound, the political science was not. Yet in response to poor political science, basic biological principles were attacked.

That evolution has occurred is as well established and accepted a scientific "fact" as is any. When such facts come under attack because of their political implications, something is seriously wrong with our process of science education; the outcome can be devastating. T. D. Lysenko, the most dominant voice in Soviet genetics from the mid-1930s to the mid-1960s, felt that evolutionary theory and Mendelian genetics had too many capitalistic overtones. In response, he demanded that all genetics research in the Soviet Union conform with Marxist ideology. In practice, this meant an acceptance of the outmoded and disproven Lamarckian concept of inheritance of acquired characteristics. The Russian biologist Zhores A. Medvedev, in his riveting book *The Rise and Fall of T. D. Lysenko,* graphically described the situation in 1948.

> Hundreds of scientists, the best and most qualified representatives of Soviet biology, were either dismissed or demoted on the basis of fabricated, slanderous, and perverted accusations of idealism, reactionary views, Morganism, Weismannism, complicity with imperialism and the bourgeoisie, Mendelism, anti-Michurinism, groveling before the West, sabotage, metaphysics, mechanism, racism, cosmopolitanism, formalism, unproductiveness, anti-Marxism, anti-Darwinism, alienation from practice, and the like. In reality these scientists were guilty of one thing only: they did not always, and in everything, agree with the ideas and hypotheses advanced by Lysenko. . . .
>
> . . . The fraudulent experiments on transformations of one species into another (wheat into rye, cultivated into wild oats and barley, cabbages into rutabagas and rape, sunflowers into strangleweed, pines into firs, etc.) were given wide publicity. Such "discoveries" were

reported by the dozen in Lysenko's journal *Agrobiologiya*, and these illiterate, shameful articles were advertised as achievements of progressive science.

The result was the loss of an entire generation of agricultural geneticists and the refocusing of research away from crop breeding toward the transmutation of species. The subsequent devastation of Russian agronomy and the Soviet wheat shortages continuing to the present day are thus Lysenko's legacy.

Antiscientific and pseudoscientific views do not occur in a vacuum. Ample rigorous sociological studies have conclusively demonstrated that astrological and other occult beliefs increase in hard times when people feel less in control of their lives. A belief in things astrological, for example, skyrocketed during the Great Depression. Recognizing this fact and writing during the onset of the double-digit inflation of the 1970s, Lamar Keene, a well-known charlatan and medium, said: "All mediums would agree . . . [that] wars, depression, personal and national disasters spell prosperity for us. The present economic stresses in the United States are good news for mediums."

At times, it seems, the social situation can get so bad that governments feel they must turn to pseudoscientific mass palliatives such as astrology to help assuage the concerns of their people. The situation in the Soviet Union in 1989 is a perfect case in point. As social unrest and ethnic violence swept through the Soviet Union and much of the Eastern bloc, and as the Soviet people began to organize against some of the devastating environmental policies that had wreaked havoc with their countryside, the Soviet government reversed its longstanding position on astrology. For years it had labeled astrology a "false science" and had prohibited astrological predictions from appearing in its publications. Nonetheless, at the beginning of 1989, *Moskovskaya Pravda*, the official publication of the Moscow Communist Party, began to run a regular astrology column. The author of the new feature was extolled as "a specialist in the arts of white, black, and other magic" and as a "master of the magical sciences."

It is an unlikely coincidence that the first officially sanctioned astrology column claimed that the conjunction of Saturn and the Year of the Snake indicated that 1989 would be a very good twelve months for the Soviet environment. This column appeared just on the heels of a growing grass-roots environmental movement and budding protests about massive Soviet environmental problems. For example, the Caspian Sea, the world's largest lake, was dangerously contaminated with massive amounts of the carcinogen phenol; the Aral Sea, once the world's fourth largest lake, had all but disappeared, the water long since diverted for shortsighted irrigation projects; and the foulness of ordinary tap water in Leningrad (now St. Petersburg) had led to long daily lines of people waiting for fresh water distribution. These and other problems were rapidly raising the environmental consciousness of the people of the Soviet Union.

Such a situation is not unique to the Soviet Union. At about the same time, Paul Kurtz, chairman of the leading national group of skeptics known as the Committee for the Scientific Investigation of Claims of the Paranormal (CSICOP), returned from a trip to the People's Republic of China and reported a marked upswing in interest in astrology and the paranormal among large numbers of Chinese. There, too, as personal living conditions deteriorated, many individuals turned to mystical explanations for solace.

Astrology, like "creation science," is a classic example of a pseudoscience. Given that absolutely no meaningful mechanisms for the influence of celestial bodies on human beings have ever been advanced, astrology clearly falls afoul of Judge Overton's first of five essential characteristics for science: it is not guided by natural law. When major superpowers tilt toward the irrationality of pseudoscience in place of the methodology and power of science, there is no telling what the worldwide consequences might be. Indeed, such a basis for decision making makes predicting the actions of governments all but impossible—a particularly troubling thought in a nuclear world.

Even thouugh the vast majority of newspapers in the United States run daily astrology columns, only a handful print a disclaimer stating that the "information" is purely for entertainment rather than predictive value. The press had a field day ridiculing Nancy and Ronald Reagan when Nancy's frequent consultations with a California astrologer became public. Subsequent reporting, however, strongly suggests that the media's stance on astrology was more a way to ridicule the Reagans than it was a principled stand against the dangers of pseudoscience.

The most disgraceful example of the media's legitimizing of pseudoscience was its coverage of another "scientific" debate, one with even greater ramifications than the first lady's personal beliefs. The media, and by that term I mean reputable entities rather than the *National Enquirer* and its emulators, paid enormous amounts of attention to Iben Browning, a business consultant with a Ph.D. in biology, who claimed that a huge, killer earthquake would occur along the New Madrid fault in southeastern Missouri on 3 December 1990. The controversy was not over the New Madrid fault itself. It is well known to have produced some of North America's largest quakes ever. To put it bluntly, there was no scientific basis for Browning's prediction of a specific quake date. Yet the media promoted what could only be called a doomsday scenario. A naive, frightened public, poorly educated in matters scientific, overreacted. Many schools in southeastern Missouri closed on the targeted day, travel agents promoted junkets to get people out of the area and reported land-office business, and many businesses were closed and shuttered.

Exactly how were the media responsible? First, and most simply, they paid attention to Browning's claims. Although he is not a geologist, Browning pushed the notion that tidal strain is responsible for earthquakes. He went on to argue that the alignment of the sun and the moon on 3 December would be especially propitious for a major quake. The National Earthquake Prediction Evaluation Council's conclusion, circulated well before 3 December, that there did not appear to be "a theoretical basis for Browning's prediction" and that, in fact, it appeared "theoretically implausible" puts him well outside the realm of science. The media might as well have promoted disaster predictions by Jeanne Dixon and others of her ilk. (Interestingly, Browning's prediction that the U.S. government would collapse in 1992 after massive crop failures was not given nearly as much attention.)

Second, the media seem to feel that journalistic integrity demands that both sides of all issues be presented, even when only one side exists. In Browning's case, like those of most pseudosciences, the situation was difficult because it was all but impossible to find reputable experts to support Browning. Undaunted by lack of expertise, the media, as they so often do, promoted whomever they could find. In this case the media's spotlight settled on David Stewart, a geologist at Southeast Missouri State University. Stewart, however, has a bit of a checkered past on earthquake predictions. In 1974, he predicted a magnitude 6+ quake for the Wilmington, North Carolina, area and demanded that the government take precautionary steps (*Science,* 26 Oct. 1990, 511). When the mainstream geologic community refused to take his prediction seriously, he resorted to paranormal means to refine his prediction, including flying over the area with a psychic. Needless to say, no quake occurred in the Wilmington area. But Stewart received a good deal of attention as an expert supporting Browning's prognostications.

Third, the media are far too willing to accept claims of previous predictions. One way the media justified taking Browning so seriously was his claim that he had accurately predicted the 1989 San Francisco quake. What he actually said, the week before the 17 October California quake, was that on or about 16 October "there [would] probably be several earthquakes around the world, Richter 6+, and [possibly] a volcano or two" (cited in *Science*, 26 Oct. 1990, 511). Given that 6+ quakes occur with an average frequency of once every three days or so, it would have been more newsworthy had Browning's prediction failed to be borne out.

Who stood to gain by promoting the prediction of a disaster on 3 December 1990? Clearly, a prediction of this sort benefited the media through increased viewers and newspaper sales. And given that Browning was hawking a ninety-nine-dollar video of himself making his pseudoscientific prediction, he was quite pleased with all the publicity.

Who stood to lose? The disruption of life in the three states adjoining the New Madrid fault should not be ignored. After I published a newspaper column attacking Browning, in the week before the supposed quake, I received a number of letters from residents in the affected area. One person wrote, "You cannot imagine the disruption of life in our communities—nor the fear caused in the very old, many of whom have even withdrawn savings from the banks fearing a catastrophe, nor among the very young who are frightened beyond belief." Even more striking, however, think about the message we are sending schoolchildren. A quack predicts a quake and we shut the schools down. Is this the nature of science in America today?

Astrology and the Browning case exemplify at least one important common theme: the public's desire to predict the future. Given the complex nature of most natural processes, scientists are often simply not capable of providing the types of predictions most people would like. An extremely fertile ground for pseudoscientists, therefore, has been their claims that go well beyond the bounds of "normal" science, especially when the predictions are related to human behavior. People are fascinated by the possibility of gaining "scientific" insight into human personality. A once popular but now discredited mechanism was exemplified by the phrenological movement that peaked in the middle of the last century. Proponents of phrenology postulated that they could tell all sorts of wondrous things about a person by feeling the shape of that person's skull. The social historian Robert Young has said that the Victorians used phrenology as "the key to all philosophical and social problems—a panacea for all social ills." With the growing acceptance of modern scientific methodology, and with the stark lack of any evidence that the claims advanced by phrenologists had any validity, the bubble soon burst on the phrenology movement, and it began to be recognized for the pseudoscience that it was.

Before we become too self-satisfied and self-congratulatory over our ability to root out pseudoscience, it is worth noting that similar modern versions of phrenology are still very much in vogue. In 1988, for example, an estimated 2 million Americans looking for work were forced to submit to a pseudoscientific test that purported to examine their suitability for employment but which Senator Edward Kennedy (D-Mass.) called "little more than a twentieth-century version of witchcraft." Conservative estimates suggest that 300,000 law-abiding citizens probably failed their tests, were assumed to be poor employment risks, and thus lost any chance of being hired by the company that tested them. According to some estimates, it is possible that three times that number of people failed their examinations. The tests in question were all administered by polygraphs, or "lie detectors," which are really nothing more than pseudoscience dressed in a high-tech facade. Senator Kennedy referred to the machines as "inaccurate instruments of intimidation"; Representative Matthew Martinez (D-Calif.) characterized a "lie detector" as a black voodoo box. In 1988, the U.S. Congress came partially to its senses with respect to this nonsense and passed the Employee Polygraph Protection Act of 1988, which almost completely bans polygraph testing in the private sector. Oddly enough, Congress refused to outlaw the use of polygraphs by the federal government—and their use in the federal sector has skyrocketed. In 1973, Uncle Sam administered some seven thousand polygraph examinations, but by 1982 that number had more than tripled to approximately twenty-three thousand governmentally administered tests. In 1988, according to conservative estimates, the frequency of testing exploded: the government alone strapped close to 150,000 Americans to these "instruments of intimidation."

More important, there is still a broad public perception that "lie detectors" are useful and accurate devices for assessing veracity. The media gave a great deal of play to the report revealed during the Clarence Thomas hearings that Anita Hill had taken and passed a polygraph exam. Not even those of us in complete acceptance of her testimony should give credence to the results of such a meaningless test.

Pseudoscience, it seems, will always fill a vacuum. As polygraph use in the private sector wanes because of federal legislation, another phrenology-like phenomenon has expanded to take its place. The use of graphology, or handwriting analysis, a pseudoscience originating in the last century, has been increasing dramatically in recent years. Graphologists, like the early phrenologists, claim that their techniques open a window onto the personality of the individual being studied. Contending, for example, that their methods allow an accurate determination of an individual's strengths, weaknesses, addictions, risk tendencies, and even health disorders, they encourage the use of graphology to determine a subject's suitability for a particular job. The truly embarrassing thing is that while it was Victorian phrenologists who made such assertions for their pseudoscience, it is modern-day graphologists who are pushing their own brand of quackery, although the evidence from scientific studies for the legitimacy of graphology is no better than that for the legitimacy of phrenology.

As ridiculous as this practice is, it is not limited to the midways of county fairs, although it is surely common there with trailer-sized machines sporting hundreds of blinking lights spitting out a personality assessment of any rube willing to shell out a few bucks. Increasingly in recent years, employers, including a distressingly large number of Fortune 500 companies, have begun turning to handwriting analysis in making hiring and promotion decisions. (Some divisions of Honeywell, Inc., and some franchises of Thrifty Rent-a-Car, for example, have taken to using graphology as a preemployment screen, and Bell Atlantic Corporation, according to a *Wall Street Journal* article, has been using it "in self-improvement exercises for managers.") It is standard practice for the analysis to be performed without ever informing the candidate. With the outlawing of the polygraph and the growing use of graphology, we must conclude that employers are simply looking for cheap and easy ways to differentiate among potential employees. Whether that differentiation is meaningful seems entirely beside the point to those relying on the tests.

Anecdotal evidence supports findings of fully controlled studies demonstrating the absurdity of buying into this type of chicanery. When identical handwriting samples have been submitted to a series of "reputable" graphologists, the range of responses are laughable. Single individuals are variously described as "very creative" and "low in creativity and imagination," "moody" and "emotionally stable," or "easygoing" and "worried and anxious." After publishing an op-ed piece on the pseudoscience of graphology, I received quite a bit of irate mail from practicing graphologists. Although virtually all defended their "science" by arguing that large samples are examined (as if using a nonsensical method to analyze a large block of handwriting is better than using that same method to examine a smaller sample), quite a few offered me "insights" into my personality based solely on my signature. A self-described "master graphologist" from Illinois volunteered: "Your scribbled signature is reminiscent of Michael before he knew how to write. It demonstrates a notable amount of what a psychoanalyst would call free-flowing-anger. It takes very little to light your fuse." The editor of the *American Handwriting Analysis Foundation Journal* said: "Now [that] I've seen your signature, I understand why you don't like graphology, since it is apparent that you don't want anyone to know you, and handwriting analysis is perceived as a threat to your anonymity." I wrote to remind her that she provided no evidence of controlled studies documenting the efficacy of her "science." The response I received included the following astounding contention: "That you did not see me 'making any claim that controlled, double-blind studies have demonstrated the efficacy of handwriting analysis' is due to the fact that I do not believe graphologists have the responsibility of proving themselves to you. Rather, you are the one who is disputing the validity of graphology, so it is up to *you* to prove it doesn't 'work.'"

This woman is partially correct—graphologists do not have the responsibility of convincing *me*. I am not the final arbiter of what is and what is not acceptable scientific practice. But she is very wide of the mark when she assumes that it is the responsibility of critics to discredit the claims of her organization. As with any scientific endeavor, the burden of proof falls most heavily on those advocating any particular hypothesis. Within the scientific framework, it is imperative for anyone advancing a hypothesis to propose and carry out a series of carefully crafted, controlled experiments. Only when such experiments yield data that are consistent with the hypothesis can we lend any credence to the claims being advanced. It takes an enormous amount of chutzpah for the proponents of graphology to make outrageous claims about their techniques, to market themselves as personnel consultants, and then to say "it is up to *you* to prove it doesn't 'work.'"

This is especially so when even some of the supposed "experts" have trouble with the methodology employed in the field. Mark Hopper, president and founder of Handwriting Resource Corporation, a Phoenix-based company specializing in graphology, asserted in an article in the 25 August 1988 *Wall Street Journal* that his company alone offers the only scientific analysis of handwriting. He argued that there is no standard for analysis because there are too many methods of interpretation, adding, "With the exception of what I'm doing, I would be opposed to all the different types of handwriting claims out there, too."

Despite widespread agreement that graphology is simply another form of pseudoscience, its practitioners are being permitted to operate in a virtual legal vacuum. No laws control who can claim to be a graphologist; the industry is entirely self-policed. Anyone interested in handwriting analysis may take a correspondence course from the International Graphoanalysis Society, the largest organization of its sort in this country, pass a test, and become certified. Alternatively, an interested person can ignore the society entirely and simply hang out a shingle proclaiming expertise.

Graphologists are not the only pseudoscientists hanging out unregulated shingles and scamming a scientifically unsophisticated and largely illiterate public. In forty-eight of our fifty states there are absolutely no checks on a growing pseudoscientific health industry: homeopathy. Only Arizona and Connecticut limit the practice of homeopathy to those individuals who have earned either an M.D. or a D.O. (Doctor of Osteopathic Medicine) degree; in every other state anyone, regardless of training, can claim to be a homeopath, hang out a shingle, and begin to collect fees. Even Stephen King, the executive director of the International Foundation for Homeopathy, finds the situation unfortunate.

Allowing homeopathy to be practiced by anyone may seem a sensible decision, since even its practitioners admit that there is no known mode of action for homeopathic remedies. King, in a telephone conversation with me, was quite open on this subject, saying that homeopathic medicine cannot work in any known physical way but might, instead, use other forces about which we know very little. The direct, physical route is ruled out for homeopathic concoctions because of the homeopath's basic dictum that "less is more." Homeopathic potions are brewed by diluting some chosen chemical in alcohol until barely a single molecule of the original chemical is left. Oddly enough, according to homeopaths, the greater the dilution, the stronger the "medicine." Homeopaths make the bizarre claim that while each successive hundred-fold dilution dramatically decreases the amount of the chemical left in solution, each dilution increases the amount of the essence of the chemical. They believe that it is the essence rather than the chemical itself which has curative power. Not surprisingly, remarkably few controlled studies have examined the powers of such elixirs, and those that have, have not found significant benefits. With no data supporting the extreme contentions of homeopathy, and with its proponents agreeing that its mode of action, whatever that might be, is against the laws of nature as we know them, homeopathy, by the criteria of Judge Overton and philosophers of science, must be ruled out of the scientific camp and into the realm of the pseudoscientific.

Nonetheless, when homeopathy was first developed by the self-educated Samuel Hahnemann at the end of the eighteenth century, it probably was, in a peculiar sort of way, an advance over the medical treatments that were then considered standard. At that time doctors practiced extremely invasive medicine; without any awareness of basic antiseptic procedure, they routinely used bloodletting, emetics, and purgatives to rid patients of a variety of symptoms. Commonly the symptoms abated, although the patients died. Such patients were honored with the epitaph "They died cured."

Against this backdrop, Hahnemann's advice to administer alcoholic concoctions of essences of chemicals rather than leeches probably saved many lives. As our understanding of disease increased and we began to find that, for example, bacterial infections respond nicely to antibiotics, the use of homeopathic "remedies" was harder to justify. A 1986 report published by the British Medical Association dismissed homeopathy as little more than witchcraft.

Yet in a society woefully scientifically and medically illiterate and increasingly populated by individuals drawn to "natural" remedies, homeopathy seems to be making a comeback. As I mentioned above, this is occurring with virtually no governmental control. Sadly, the victims are often those most in despair, the aged and incurably ill, who do not deserve to have their hopes needlessly raised and then dashed. Morally we are all responsible for the disgraceful lack of governmental control.

Similarly, as a result of a clever Medicare scam, we are also monetarily responsible.

Consider the case of my father. Before his death, he was confined to a wheelchair with a debilitating, deteriorative neuromuscular condition of unknown origin. A host of doctors at some of the country's best hospitals were unable to offer either diagnosis or hope. Although I postulated that his condition might be due to DDT poisoning, since he was regularly covered with the chemical during his service in the marines and the onset of his condition began after a crash diet, which might have caused DDT to be released from storage in his fat cells, no one cared to pursue the matter. Finally, my parents decided to look for alternative health care. Living in an area of Florida well populated by the elderly and infirm, they easily came up with a list of homeopaths. They chose the closest one and immediately paid her a visit. She was extremely excited about the DDT hypothesis, and, jumping on another of Samuel Hahnemann's eighteenth-century ideas, "like cures like," she prescribed DDT pills for my father.

You can imagine my distress upon learning of my father's treatment. In speaking with the homeopath, I quickly discovered that she had no formal training in medicine or homeopathy but had, in her own words, "worked with a master of the field." She assured me that her treatment was absolutely safe because the pills she gave my father were so dilute that they did not have a trace of DDT in them, only the essence of DDT. She went on to say that as he began to improve, she would change the pills to an even more dilute version, one with even more essence, that is, even less DDT than the pills that had no DDT before! She was worried that too much essence too early in his treatment might well cause complications.

My discussions with state and federal medical regulatory authorities all proceeded in the same fashion. Everyone was shocked and distressed to hear that DDT in any form was being prescribed as a "medication," all acknowledged that DDT was not on any approved list for such uses, but none was willing to consider taking any action.

My father swallowed months' worth of the pills along with the homeopath's line. Although no physical harm came from the treatment, his spirits were lowered even further when no positive effects were seen, and it was difficult to get him to see any doctor subsequently.

An equally alarming part of the affair is that Medicare picked up the tab for my father's "treatment." The homeopath, you see, works out of an office run by an M.D. who signs all of the paperwork, which claims that he has seen each patient. The federal government is happy to pay for such "consultations." Needless to say, my father never saw, much less spoke with, that doctor. Ambrose Bierce was not far off when, in 1926 in *The Devil's Dictionary*, he defined the homeopath as "the humorist of the medical profession." I am afraid, however, that under the present circumstances the homeopath is laughing all the way to the bank.

Gullible and desperate Americans are seemingly always searching for easy answers to complex problems, even when no scientific data support the fantastic claims advanced by proponents of pseudoscience. The fad diet industry, for example, has become a huge business. People jump at the opportunity to plunk down their dollars for pills that will "melt the fat away" in the complete absence of dieting or exercise, unmindful that these claims are scientifically absurd and even publicly ridiculed. Unfortunately for many of us, weight loss, like all difficult personal decisions, requires hard work and time, for which pseudoscientific shortcuts are simply not adequate substitutes.

The production of subliminal self-help audiotapes is another industry that has been capitalizing on the same pseudoscientific, quick-fix mentality. These tapes supposedly include hidden messages masked by soothing music or nature sounds. The messages are supposed to work wonders while the consumers need do nothing but sit back and relax. The tapes are designed to "cure" all sorts of human frailties. Manufacturers claim that some will help "listeners" lose weight and stop smoking; others are supposed to treat sexual dysfunction, constipation, high blood pressure, and poor study skills. Although many people might well dismiss subliminal self-help tapes out of hand, the tapes are increasing in popularity; in 1986, the last year for which data are currently available, they represented an industry of approximately $50 million annually.

Although each tape comes with strong assurances of success and ample testimonials from users, there have, until recently, been no controlled, scientific studies examining whether or not the tapes work. A number of scientists have now published the results of such studies in peer-reviewed journals, and the results are striking: the music and the nature sounds might be relaxing, but the subliminal messages are useless.

By far the best study was undertaken by Anthony G. Greenwald, a psychologist at the University of Washington. Along with his three co-workers, he examined the effects of tapes designed to improve memory and those designed to improve self-esteem. After being subjected to two pretests to evaluate their starting levels, subjects were each given a single tape, labeled either memory or self-esteem, and directed in its use. Subjects were not told that the researchers had switched the labels on half of the tapes, so half of the subjects who thought that they were listening to subliminal self-esteem messages actually had been given memory messages and vice versa. At the end of a month each subject was again tested. The conclusions reached by Professor Greenwald and his colleagues, published in 1991 in *Psychological Science,* a professional journal put out by the American Psychological Society, are impossible to misinterpret: "After a month of use, neither the memory nor the self-esteem tapes produced their claimed effects" (vol. 2, no. 2, 119–22). And, bizarrely enough, the memory tapes seemed to do a better job of increasing self-esteem than did those created to do exactly that.

Professor Greenwald likens the manufacturers of these useless products to snake oil salesmen of old. Some are clearly unscrupulous, well aware that their products are without redeeming value, but others, he contends, believe that the gullibility of the public can lead to self-cure. In other words, if a buyer believes that a memory tape will improve his or her memory, then that belief alone might enhance memory performance. Indeed, the published results indicate that tape users did feel that they were better off after using the tapes. Unfortunately, none of the objective tests designed to measure improvement demonstrated any positive results. Thus, although people thought their memories were much improved, the evidence indicated no such improvement.

Having a scientific outlook means being willing to divest yourself of a pet hypothesis, whether it relates to easy self-help improvements, homeopathy, graphology, spontaneous generation, or any other concept, when the data produced by a carefully designed experiment contradict that hypothesis. Retaining a belief in a hypothesis that cannot be supported by data is the hallmark of both the pseudoscientist and the fanatic. Often the more deeply held the hypothesis, the more reactionary is the response to nonsupportive data.

An example of nonsupportive data which angered a good number of people by forcing them to reexamine a deeply held belief comes from the social rather than the natural sciences. The hypothesis was that "capital punishment deters murder," and the study was performed by sociologists William Bailey and Ruth Peterson, of Cleveland State and Ohio State Universities respectively, and published in 1989 in the *American Sociological Review* (vol. 54, 722–43). In a painstaking and massive analysis of every instance of capital punishment in the United States between 1940 and 1986, Bailey and Peterson examined homicide records to determine whether murder frequencies decline significantly as a function of the frequency of capital punishment. They looked for both short-term deterrence, during the month in which the execution was carried out, and longer-term deterrence, during subsequent months, and found neither. Their conclusions hardly require a Ph.D. in sociology or criminology to understand: "The cumulative effect of capital punishment on homicides during the [month of the] execution and subsequent months is essentially zero" (722).

Bailey and Peterson then took their analysis one step further. Reasoning that the deterrent effect of state mandated executions might vary as a function of the publicity that those executions received, the two sociologists rated the publicity associated with each as high, moderate, or low. Again, their conclusions should be easily interpretable by anyone: "No evidence for 1940–1986 supports the conclusion that highly (or moderately) publicized executions significantly deterred homicides" (736).

One might hypothesize that although publicity of executions is widespread, deterrence should be noted only in those states that practice capital punishment. It was widely known at the time that Gary Gilmore was executed in Utah on 17 January 1977, but this fact surely might not have had a deterrent effect in any of the thirteen states in which capital punishment was not a legal option. Bailey and Peterson examined exactly this point by looking at homicide frequencies in states with and without the death penalty. As before, their results are impossible to misinterpret: "Residents of jurisdictions with capital punishment ran a slightly, but significantly, greater risk of being killed during the period" (736). Rather than supporting the hypothesis, the data in this instance demonstrate the reverse; instead of being safer, people living in capital punishment states were actually more likely to be murdered than were citizens living in those states that did not execute criminals.

The Bailey and Peterson study is a particularly good example of the way in which hypothesis testing should proceed. The two sociologists began with the basic hypothesis and put it in a form that was falsifiable. When the available data were inconsistent with that hypothesis, they refined it, again in a form that was falsifiable. After reiterating the process, they were left with the conclusion that the available data simply do not support the original hypothesis that capital punishment deters murder. That original hypothesis now needs to be discarded.

The United States is alone among the countries of the Western world in levying capital punishment. Since 1976, when the U.S. Supreme Court ruled that such punishment is legal, the states have certainly sentenced large numbers of people to death. There are currently more than two thousand people on death rows around the country, the annual number of individuals executed each year has grown steadily throughout the 1980s and early 1990s, and well more than two hundred people have been executed in the eighteen years since Gary Gilmore faced a firing squad.

People favor the death penalty for a wide variety of reasons, but the most commonly voiced argument seems to be deterrence. Proponents, however, now need to acknowledge that the deterrence hypothesis has been soundly refuted by the available data. Although it is possible that most capital punishment advocates may not change their minds because of the study performed by Bailey and Peterson, they need to be honest enough to alter their reasons for advocacy. If, for example, they state quite plainly that they favor the death sentence for reasons of revenge and retribution, they may be criticized for being morally out of step with the rest of the Western world, but no empirical study will prove them wrong.

Even worse than those who refuse to back away from hypotheses that are contradicted by data are the growing number of "new age" mystics who emphasize that their strange notions are *un*supported by science. Peter Tompkins and Christopher Bird, in their book *Secrets of the Soil*, are masters of just this method of argumentation. Throughout their book Tompkins and Bird are blatantly disdainful of modern science, repeatedly and gleefully pointing out that science has been unable to verify their claims (or even understand the mechanisms proposed), implying that such lack of verification only enhances their conclusions:

— "It is a fact," they claim, that there are "places where cars mysteriously roll backward up a hill as gravity turns to levity."

— They ask readers to take seriously "the many stories about trees which had borne no flowers or fruits for years, suddenly blossoming and bearing when threatened with an axe or a chain saw."

— Or they turn to basic, old-fashioned alchemy with their claims that plants (and a few knowledgeable people) can transmutate the elements, changing, for example, extra potassium into needed sulfur.

The most distressing aspect about all of this is that Tompkins and Bird represent a growing anti-intellectual movement. Although that "new age" movement recognizes many of our most pressing environmental problems, its pseudoscientific solutions leave an enormous amount to be desired. It is absolutely true, as many "new agers" point out, for example, that we would all be much better off if we turned to organic agriculture, if we seriously practiced soil conservation, and if we began to plant multiple crops instead of monocultures. But it doesn't serve any productive purpose to turn to nonscientific, quick-fix solutions that draw heavily on astrology and homeopathy. As Tompkins and Bird point out, a growing number of people are turning to "bio-dynamic" agriculture, which promises that farmers and gardeners will increase crop yields, stymie insects, and enjoy healthy, productive lives well into their second century. Oddly enough, that all the original proponents of this blather died before reaching one hundred, including the founding father of "biodynamic" agriculture, Rudolf Steiner, dead at a youthful sixty-four years, does not seem to give adherents second thoughts.

No, we have not made environmental strides when the proponents of "biodynamic" agriculture convince large numbers of us that a homeopathic amount of a concoction called BD 500 (mostly cow manure stored for a year underground in a cow horn), mixed with rainwater and stirred for an hour with a wooden stick, imbibes the energy of the cosmos and, when sprinkled over crops, elicits wondrous growth. Nor are we better off if so many of us naively believe that magnificent growth can be obtained by another homeopathic elixir called sonic bloom dripped onto hungry plants awakened by just the right music. We might even be worse off if, instead of taking meaningful political or environmentally remedial action, environmentally concerned citizens decide to burn a blend of cow manure, ghee, rice, and sandalwood in an inverted copper pyramid because they've been led to believe that igniting such a mixture will significantly reduce the amount of pollution in the world.

Consistently, the claims of pseudoscientists are more attractive than those of scientists. Typically, pseudoscientists promise to accomplish more in a shorter time than any scientist could ever imagine. And nothing seems to be impossible to a pseudoscientist. Given all of this, it should not be terribly surprising that so many of us turn so eagerly toward the facile solutions of pseudoscience. However, we simply cannot afford to take the easy way out and buy into false hopes and promises. We must approach any and all claims with a healthy dose of skepticism, asking how those claims are consistent with natural laws and consistent with empirically derived data, and we must be sure that all hypotheses advanced are falsifiable. Only then will we be in a strong position to evaluate our understanding of the natural world.

Chapter Three

⟋ン

Anecdote, Coincidence, and Pattern

Understanding
the Language
of Science

More than two hundred years ago, in *The Wealth of Nations,* economist Adam Smith wrote that "science is the great antidote to the poison of enthusiasm and superstition." But even now in our modern,
technologically advanced world, many people often have difficulty
differentiating science from superstition. As we've seen in the previous
chapter, this should not be the case, since most philosophers of science
agree that the most important distinguishing criterion of real science
is its use of falsifiable predictions; that is, real science makes specific
predictions that, if incorrect, could conceivably be disproved by carefully designed experimentation. Superstition and pseudoscience, on the
other hand, offer no such predictions and hence close themselves
soundly against debate or disproof. Instead, they profess to tell absolutely how the world is or was.

The philosophers of science have clearly had an impact, and the
savvy purveyors of pseudoscience have come to recognize the critically
important role that falsifiability plays within the scientific community.
Accordingly, they have begun to wrap their ideas in a veneer that gives
lip service to the concept, and they have begun to try to pass off generic
prophecy as specific prediction. It is not possible, therefore, to turn to
"creation scientists," astrologers, or others of their ilk and ask simply:

"Are your ideas testable, are they falsifiable?" expecting an honest denial and a retreat from their claims. Instead, in every case, we will be told: "Of course they are. They're scientific."

Unfortunately, we must always be on our guard and work hard to see through such simple declarations. Our goal should be to reach a level of sophistication sufficient to enable us to examine the specific hypotheses or types of claims made by those who, either purposefully or naively, are advancing anti-intellectual views. In other words, it is essential that scientific methodology be well enough understood so that all of us can distinguish truly testable predictions from those that are purposefully designed to fool us into thinking that they are testable. Furthermore, it is not even enough just to know when a prediction is falsifiable. It is imperative to determine whether a particular idea has actually been proven false. Although differentiating between coincidence and repeatable, verifiable pattern is often a difficult task, just that sort of distinction is necessary for rational decision making. To accomplish all of the above requires at least a rudimentary understanding and appreciation of experimental design and statistics. Both are essential if scientific advances are to complement and not contradict the needs of society, and, more important in many ways, both are prerequisites if the population at large is to be an active participant in issues of science and public policy.

At this time, our society is largely illiterate mathematically. Because science is so often taught as a collection of facts rather than as a dynamic process, it is no wonder that so many of us are so readily taken in by charlatans. But this is not the way things have to be. Neither scientific methodology nor a rudimentary appreciation of statistics has to remain inaccessible to the general public.

Let me begin by discussing the most common problem associated with experimental design. Bias can creep into the work of even the best-intentioned scientist. Although unintentional bias is absolutely distinct from fraud, its consequences can be similar as well as quite serious. For this reason, the best experimental design is one that is set up in such a way that the person or persons collecting data from an experimental and a control group are unaware which is which. (A control group is one upon which no experimental manipulation of any sort is performed. By comparing appropriately matched control and experimental groups, it is possible to determine the effect of the experimental manipulation.) Such an experiment is called a blind experiment and dramatically reduces the effects of any experimenter bias that might exist.

In many situations, a double-blind experiment is even better than a single-blind one. Consider a situation in which a new drug is being tested. An appropriate experimental design would call for the establishment of two groups: an experimental group consisting of people with a particular disease who will be given the experimental medication, and a control group also comprising individuals with the disease who will not be given the experimental drug. Most people would immediately recognize a problem with what has been outlined so far; the health of the people receiving the new drug might improve, not because of the drug itself, but because their hopes and spirits are raised by the mere fact that they are participating in a study and are being given medication. The simple answer is to tell individuals in both groups about the study but not whether they are in the control or the experimental group, that is, not let them know if the dosage they are taking is an experimental drug or a placebo. Under such conditions, the patients themselves will not bias the results.

Of course, a second level of bias is still possible. Some doctor, or team of doctors, will have to examine each patient. If the doctor knows which patients are actually taking the experimental medication and which are not, she might unconsciously evaluate the patients slightly differently. Again, the solution is that this level of the experiment also be conducted in a blind fashion, that the doctor remain purposefully ignorant about which patient is which. Not until the experiment has been concluded should she learn which patients were on medication and which were ingesting placebos. Unfortunately, far too frequently, especially in medicine, such experimental controls are simply not in place. Without such controls, a doctor might well know more than is necessary and thus open her interpretations of results to unintentional bias. When this occurs, it is difficult to have confidence in the experiment's conclusions. A personal experience nicely illustrates this claim.

Beginning at around one year of age, my oldest son, Alex, was extremely short for his age. His pediatrician, concerned that Alex might have a growth disorder, recommended that we determine Alex's bone age. The test is simple. The child's wrist is x-rayed, and a trained radiologist decides with what chronological age the bones in the wrist correspond. If the bone age matches the chronological age, the pattern of growth is normal. If the bone age appears much younger than the chronological age, then some other problem is indicated.

We went through the process and dutifully received a report indicating a perfect match between bone age and chronological age. It wasn't until a year or so later that I managed a look at my son's records and discovered the sheet requesting services from the radiologist. The pediatrician was required to fill in the chronological age of the child on that sheet before the radiologist looked at the x-rays. When I discussed the issue with the doctor, he readily explained the process. The x-rays are routinely submitted to the radiologist along with the patient's chronological age. The radiologist then looks up x-rays of wrists of that age in a standard radiological atlas and decides whether or not the match is good. In Alex's case the radiologist found the match to be close.

There is absolutely no question that this procedure saves an enormous amount of the radiologist's time. But the methodology is atrocious. Can there be any doubt that such an experimental design is apt to bias the conclusions reached in favor of a relatively close match between bone and chronological ages? Because the radiologist was given more information than was necessary, the conclusions were compromised.

When care is taken to design an experiment correctly, the results are much easier to interpret. But careful interpretation is still absolutely necessary because the results may not always be what they so clearly seem to be. Consider the case of drug testing. Those individuals leading the fight in our war against drugs seem to have lost sight of a basic statistical principle. Quite simply, even under the best of circumstances, any randomly administered drug test is going to turn out positive for a number of individuals who have never used drugs. Although everyone might recognize the validity of that statement, few people recognize the magnitude of the problem.

Even though statistical arguments sometimes scare people away, I am going to risk one here. The numbers are not all that difficult to understand, and a meaningful public drug policy can be achieved only if we appreciate those numbers. In addition, the broader lessons concerning data interpretation are striking.

Consider the following situation: As the personnel manager of a large firm, you decide to subject every job applicant to a urinalysis for a range of illegal drugs. As a matter of policy you refuse to hire anyone who tests positive. Is this a rational policy? In other words, are you confident that those individuals testing positive really are likely to have been using drugs? An answer to this question, obviously, has to begin with the accuracy of the test. Let's be generous and say that the test is as accurate as any on the market (and much more accurate than most); it determines drug users with 99 percent accuracy. The false positive frequency, the frequency with which someone incorrectly tests positive, is therefore only 1 percent. With such a wonderful test at your disposal, how can you go wrong?

As amazing as it seems, even under such circumstances anyone who tests positive is still most likely not a drug user. The exact odds depend on the percentage of the total population using the drugs for which you are testing. If, for example, only 1 person out of every 1,000 job applicants is a drug user, a test with a 1 percent false positive frequency will yield 11 positive cases: 10 false positives and 1 true positive. Since you cannot distinguish the false positives from the true positives, you can conclude only that each job applicant who tests positive has about a 9 percent chance (1 out of 11) of actually being a drug user. If 1 in 500 job applicants uses the drugs being tested, your positive applicant has about a 17 percent chance of being a user. Even if 1 in 100 applicants uses drugs, your positive applicant still only has a 50 percent chance of being a drug user. Given those odds, you would be turning away an awfully large number of perfectly competent, drug-free prospective employees for absolutely no reason.

Let's leave our dream world and recognize that in the real world two factors might be quite different. First, the false positive frequency is typically significantly greater than 1 percent. Second, the actual number of drug users might be significantly greater than 1 in 100. Because these two factors are working in opposite directions, in the real world a positive test result is practically meaningless. For example, for a test with 90 percent accuracy and a population in which 1 in 10 used the drug for which the test was being run, a positive applicant still only has a 50 percent chance of actually being a drug user.

These statistics do not mean that drug tests are worthless. Results are much easier to interpret when the tests are administered only to individuals who have manifested drug-induced behavior and when all positive tests are repeated. When a single test is given indiscriminately to all employees or to all job applicants, however, the results are impossible to interpret meaningfully.

These facts make the governmental frenzy toward wholesale drug testing particularly problematic. A while back, then Attorney General Edwin Meese said that he "would like to see the day where every person arrested is subjected to a urine test." He also called for widespread testing in the workplace, claiming that "we need to have drug testing in most areas of work." While few would dare argue publicly with his belief that the nation needs "zero tolerance of drugs in any place, any time," many of us know that such sweeping use of indiscriminate testing will implicate staggering numbers of innocent people.

Nonetheless, the business community is placing increasing reliance on just such indiscriminate testing. Among Fortune 500 companies, on-the-job drug testing and preemployment screening are increasing rapidly. A number of legislators, recognizing these problems, have demanded that employees not be fired on the basis of a single positive drug test. But drug tests are expensive, and no such protection yet exists for job applicants. The policies underlying the war on drugs cannot be considered humane when so many innocent people become casualties. A more sophisticated understanding of statistical data would yield much more rational and much more effective policy.

The drug-testing scenario makes it clear that interpretation is a critical portion of experimental design. Under the best of circumstances, random drug testing (or indiscriminate screening for any disease) will complicate the lives of a huge number of perfectly healthy people. In the absence of public awareness of such a phenomenon, numerous lives and careers will be unnecessarily tarnished.

While statistical arguments of this sort are not necessarily intuitively obvious, neither can they be ignored. They enable us to differentiate mere coincidence from actual pattern. The need to make such distinctions is critical and forces us to ask ourselves difficult but probing questions such as: When should we be surprised that a particular series of events has occurred, and, therefore, when do we need to look for the cause of those events? Alternatively, when should we just accept that a particular series of events is simply the expected outcome of random chance?

As an illustrative example, take the following situation. Your child comes home from kindergarten announcing that two people in her class share the same birthday. How surprised are you? How surprised should you actually be? Should you assume that some mystical force was at work to create such an odd result? Well, if your child is enrolled in a kindergarten class with 22 other children, the odds are greater than

50 percent that two kids will share a birthday. Although this result might seem counterintuitive, the statistical explanation is really quite simple. Even though there are 365 possible birthdays in the year and only 23 kids present, there are so many ways to match up pairs of kids that the odds favor some set of two having the same birthday. Your daughter, for example, could have the same birthday as any one of the other 22 remaining students, a second student could pair up with any one of 21 other students, a third student could pair up with anyone of 20 other students, and so on. In a class of 23 kids there are 253 distinct pairwise combinations of birth dates possible. In fact, if two classes of this size were combined, you really should be surprised if no two kids shared a birth date; given the number of possible combinations (1,035), there is approximately a 95 percent probability of a match. Even with so many combinations possible, the odds of a match are not 100 percent because only 46 people (or the possibility of 46 distinct birthdays) are represented. We cannot be 100 percent certain of a match until 366 kids are present.

An appreciation of probability might impact favorably on some of the personal decisions we make. Large numbers of us, for example, play state-sponsored lotteries without actually understanding the statistical odds of winning. Again, the misunderstanding points to a deficit in mathematical and probabilistic thinking which goes well beyond the specific examples discussed here.

In addition to playing the lottery regularly, a surprising number of people invest money in one of those tip sheets in newspaper ads promising to sell you winning numbers. You know the ads I mean—the ones that say, "If you want to win the Ohio Lottery more often, start using our 'HOTSHEET,'" and "If you are tired of collecting those 'losing' tickets, get our 'Ohio Hotsheet' today!" In case you're one of the people attracted to these publications, let me ask you a few questions before you spend your money. These questions are designed to help you find out just how much you really understand about the odds of winning the lottery and to persuade you not to waste your money on pseudoscientific prognostications. We'll use the Ohio Super-Lotto as a model. The game is run twice weekly with a minimum jackpot of $3 million. To win it all, you must correctly select the six numbers between 1 and 40 which turn up at a particular drawing.

Let's start out simply. *What are your chances of actually winning the jackpot, of having all six numbers correct, if you purchase just one ticket?* The odds are clearly against you: you have 1 chance in 7,059,052.

Which set of numbers is more likely to win: 1, 2, 3, 4, 5, 6 or 3, 12, 17, 23, 29, 36? They are both equally probable; the first set just seems odder because the numbers form a nice sequence. Any set of six numbers has exactly the same probability of turning up as does any other set of six numbers.

On Saturday, your friend always bets on the numbers that won on Wednesday. Is this a crazy thing to do? Isn't it twice as unlikely for the same six numbers to turn up in two drawings in a row? Although you might be surprised to learn it, your friend has exactly the same odds of winning as does anyone else. Again, any six numbers are as likely to show up as are any other six numbers, regardless of how many times they've shown up before.

If a number hasn't shown up for months, is it due? Absolutely not. Every time a ball pops into the chute, it is an independent event. There is no way, unless the lotto game is rigged, for a number's history to affect whether or not it turns up in a particular drawing, even if it hasn't shown up in ten years.

What about using the auto-lotto option? Isn't it twice as hard to win if you let the computer select your numbers because then those numbers have to turn up twice in the same week? Nope. Whether you choose the numbers or the computer picks them makes no difference; the odds are identical.

The message is quite simple: any number or combination of numbers is every bit as likely to win as is any other number or combination of numbers. No numbers can be more likely than any others in any given week. There is no information that anyone can sell you that will increase your chances of winning. The lottery tip sheets are absolutely useless; betting the numbers that are provided will not increase your chances of winning in the least.

Why then, you ask, do these "services" claim to produce so many winning numbers? The answer is in their ads: "Telling our readers to play more than six numbers is the key to our success." Instead of giving you only six numbers, the tip sheets generously provide you with ten numbers to bet. Of course you're more likely to hit correct numbers when you can choose ten instead of only six. The ads fail to tell you two things, however. First, their ten numbers are absolutely no better than any ten numbers you might come up with on your own. Second, if you want to take full advantage of the ten numbers that they provide and be sure that all combinations of six are covered, you would have to buy 210 separate tickets. And even if you spend the $210, your odds of actually winning the jackpot will still be substantially less than one in seven million.

Let me pose two more questions. *First, if the people selling the lottery tip sheets are so confident of their own numbers, why don't they stop wasting their time selling the things and simply win the jackpot twice each week?* Think about it. With a minimum jackpot of $3 million dollars twice each week and approximately eight drawings each month, these people would have to sell 2.4 million copies of their $10 tip sheets each month to make as much money as they would by investing $1.00 on each superlotto drawing. Finally, *If they don't have enough confidence in their "information to put it to work themselves," why should you?*

While I would not argue that it is important for everyone to have an intimate understanding of the mathematics behind state lotteries, I do think that ignorance of this sort is indicative of numerical illiteracy, and such illiteracy can have grave consequences.

Again, a couple of medical examples will make my point. Quite a few years ago the word was out that laetrile, an extract of peach pits, worked wonders on cancer. Testimonials abounded from individuals who ingested the stuff and were supposedly cured. The press jumped on the bandwagon, and the United States Food and Drug Administration (FDA) was castigated for failing to approve the use of laetrile. People flocked across the border to Mexico where the extract could be legally prescribed. Did the FDA act correctly? Absolutely. In the absence of compelling evidence from controlled studies, it is unwise and possibly dangerous to make public policy on the basis of anecdotes. Perhaps many people ingesting laetrile did recover. Does that fact alone mean that laetrile was the cause of their good fortune? Of course not, because far too many questions are left unanswered by those anecdotes. How many other patients were left unaffected or were actually harmed by the peach extract? What other characteristics differentiated the laetrile takers from the nonusers? How did their diets differ? What is the normal frequency of individuals going into spontaneous remission? How long did the "cure" last? The list of questions could go on virtually forever. In fact, scientific studies were conducted as far back as 1953, when the Cancer Commission of the California Medical Society concluded that laetrile was worthless as a treatment. Although all subsequent studies yielded similar results, they stopped neither the unscrupulous from pushing laetrile nor the desperately ill from accepting it.

It is easy, especially when the stakes are high, to accept anecdotes in lieu of scientific patterns. And the more articulate and well educated the spokesperson, the more easily we find ourselves leaping to misplaced hopes and erroneous conclusions. The amazing and immediate success of Norman Cousins is the classic example. Remember that Cousins was desperately ill with an undiagnosable ailment. After numerous specialists had given up on him, Cousins took control of his own treatment. He prescribed laughter and, accordingly, spent hours watching classic comedy films. Cousins recovered. He went on to promote his view of medicine in a best-selling book, numerous talk shows, and countless interviews. He was even given a professorship at a medical school. My point is not to argue that laughter is useless as a treatment. Rather, I would argue that Cousins' cure represents nothing more than a single anecdote. Further, all of the testimonials that he might be able to gather represent nothing more than a collection of anecdotes. Anecdotes, however many may be amassed, can never constitute pattern. A meaningful pattern can only be recognized by using appropriate scientific methodology and by being aware of the basic rules of probability. Again, this does not mean that Cousins is wrong but only that in the absence of controlled studies, it would be premature and dangerous to formulate public policy based on his observations. Although at the personal level it is perfectly understandable that desperately ill people might turn to unproven methods (especially classic comedy films) and hope for recovery, society as a whole cannot afford to make such decisions.

Unfortunately, some of my research (done collaboratively with Jeff Witmer, a statistician at Oberlin College) has demonstrated that a large segment of the medical community is as unable to differentiate between anecdote and pattern as is the general public. When religion gets mixed together with medicine, reason seems to be the loser. Not only are too many doctors willing to accept religious anecdotes as evidence of medical efficacy, they are downright hostile to calls for controlled studies designed to prove whether or not a particular treatment is effective.

Increasingly, in recent years religion, medicine, and the law have come into conflict. A wide range of states (Florida, Ohio, Massachusetts, and California, to name a few) have brought charges of child endangerment and neglect against parents who, claiming religious proscriptions, have withheld medical treatment from ill children, thus allowing (or precipitating) the child's death. The government has been largely successful in winning convictions against such faith healers.

Against this backdrop, an astonishing report was published in 1988 in the *Southern Medical Journal*. Randolph Byrd wrote a paper claiming that intercessory prayer, that is, prayer on behalf of another person, was instrumental in improving the recovery of patients in the coronary care unit of San Francisco General Medical Center. Byrd claimed that patients who, unbeknownst to themselves and their doctors, received the prayers of specially designated church groups suffered fewer complications than patients for whom no additional prayers were said. If Byrd's results were scientifically accurate, they would be of the utmost importance.

As a large number of scientists, doctors, and statisticians have pointed out, Byrd's work suffers from a wealth of crippling methodological flaws. Statistically, Byrd's paper is unsatisfactory because it did not treat patients as individual entities. In a multivariate analysis of complications arising after hospital admittance, Byrd found that his experimental subjects suffered statistically fewer complications than did his controls. However, many of the complications listed are not independent of one another, so that a single patient was more likely to manifest a range of symptoms rather than just one. Because of this problem Byrd's analysis is not persuasive, and there is no way to assess the efficacy of the prayers. Procedurally, Byrd reported that the intercessors were given pertinent updates on the conditions of the patients for whom they were praying. Such interim reports are at odds with the usual double-blind protocol. In short, his work cannot be said to support the efficacy of prayer.

However, Byrd's flawed work did prompt us to look more closely into the field of prayer and medicine. What we found was far from encouraging. The published material, though quite limited, is unequivocal. We were able to turn up only three published reports detailing a scientific analysis of either prayer or faith healing. A 1965 study (*Journal of Chronic Diseases* 18, 367–77) found prayer to play no significant role in treatment of rheumatic disease, and a 1969 report (*Medical Times* 97, 201–4) found prayer to be ineffective in treating childhood leukemia. Finally, a 1989 paper in the *Journal of the American Medical Association* (vol. 262, 1657–58) demonstrated that graduates of Principia College, a Christian Science College in Illinois, suffered significantly higher death rates than did a control group of graduates from the University of Kansas.

Because we thought it possible that medical journals would shy away from publishing studies examining the efficacy of intercessory prayer as a medical treatment, we sent questionnaires to the editors of thirty-eight medical journals asking if they had received, but not published, reports on this topic. All responding editors indicated that they had not received such reports. We also asked the editors if they would publish a letter requesting information from their readers on this topic.

Many letters were published. Immediately our phones began to ring and our mail boxes to overflow. Remember, we asked doctors to contact us with any information they had on scientific studies examining the utility of prayer as a medical treatment. What we received were testimonials, anecdotes, and wishful thinking. Numerous doctors told us dogmatically that prayer had worked on their patients and loved ones. When we asked for scientific verification that it was prayer that made the difference, most were upset by our skepticism; none provided any independent documentation. Other doctors told us that they were aware of someone who had used the scientific method to demonstrate the efficacy of prayer. We tracked down every lead of this sort and, amusingly, often ended up with someone suggesting that it was our original informant who had done the critical study. The experience was reminiscent of a perverse version of the children's game of telephone, in which a string of doctors, all absolutely sure that someone else had conclusively demonstrated, with impeccable scientific methodology, the efficacy of prayer, left us with nothing more than a few garbled references to nonexistent studies.

The plain and simple fact is that no one using anything akin to scientific methodology has yet successfully demonstrated that intercessory prayer is an effective medical treatment. This result will surely not prove surprising to most religious leaders, who I am sure would find it hard to believe that any deity would act on behalf of the sick only after numerous prayers were offered by third parties.

What I find most disquieting about our results is that such a large number of doctors are willing to accept anecdotal in lieu of scientific evidence. A prayer is offered, a patient is cured, and many doctors believe that the former caused the latter. This combination of acceptance of anecdote and apparent lack of skepticism is dangerous stuff. It may, in large part, be responsible for the tragic spate of childhood "faith-healing" deaths, for our willingness to accept bizarre remedies such as laetrile and homeopathy, and for the growing clamor for the FDA to release a wide variety of untested and unproven drugs. When the public is both ignorant of scientific methodology and generally unskeptical, we have a serious problem. When the medical community is ignorant and unskeptical, we have a crisis. It is time that our medical schools stepped forward and began to address this point head-on.

A sophisticated understanding of experimental design and probability is essential because it keeps us from leaping to intuitively obvious or spiritually satisfying conclusions that are factually incorrect. Without such an understanding, it is temptingly easy to be taken in by those who would use the language of science to advance their own partisan beliefs. Faith healers and pseudoscientists are not alone in this endeavor. Representatives of many industries have also been known to misuse science in an attempt to increase their own profits. One such example is the purposeful confusion about the difference between information and interpretation. In an ideal world, the former should be value free and shared as widely as possible. In such a world, it is the interpretation of data, rather than the data themselves, which engenders lively scientific debate. Unfortunately, we do not live in such an ideal world.

The problems swirling around federal legislation dealing with carcinogens are an excellent example of how data are intentionally politicized. In 1978, Congress passed a law requiring that the secretary of the Department of Health, Education, and Welfare publish an annual report containing "a list of all substances (i) which either are known to be carcinogens or which may reasonably be anticipated to be carcinogens and (ii) to which a significant number of persons residing in the United States are exposed." The idea, which was rather straightforward, was to quickly inform the public and health professionals about the existence of carcinogenic hazards in the environment. Because the reports were to be compiled by government scientists neither funded by nor obligated to private interests, the annual reports would be a compendium of objective information free of all political pressure and thus would provide the data necessary to set meaningful priorities concerning exposure reduction regulations and future research.

After a general restructuring of cabinet departments, the First Annual Report on Carcinogens was released in July 1980 by the Department of Health and Human Services (DHHS). Not surprisingly, the Second Annual Report was released in 1981, and the Third Annual Report was published a year later in 1982. Then things started to go awry. The Fourth Annual Report, due out in 1983, was not available until 1985, and the Fifth Annual Report, originally scheduled for 1984, was finally produced in 1990 after a major federal lawsuit. The delays in actual publication stemmed from both governmental inefficiency and industry attacks. Jacqueline M. Warren, attorney for the Natural Resources Defense Council, succinctly explained the larger problem when testifying before a DHHS committee: "A horde of industry lawyers concerned about the impact of the Report on liability, balance-sheets, and possible future regulation, has descended upon the Department, intent on transforming the Annual Report into a highly political, contested document."

The Chemical Manufacturers Association (CMA), for example, first distributed a position paper calling for dramatic changes in the way the report is produced. Perhaps their most striking modification is the demand that drafts of the report be made available for public comment prior to publication. Public comment and the inevitable legal challenges to specific items would likely delay publication of all subsequent editions for many additional years. In addition, should any of those legal challenges prove successful, the very nature of the report would change. Second, the CMA brought suit in federal district court in Louisiana looking for an injunction limiting publication altogether. Happily, the government's right to publish such useful environmental data was upheld.

In its blind desire to protect its pocketbook, the chemical industry was confusing information with regulation. The Annual Reports on Carcinogens are not meant to be definitive statements about the toxicity of any individual chemical. Rather, they provide the government with the opportunity to collect and summarize at timely intervals all available data on the carcinogenicity of various compounds. The reports are thus a unique reference tool. Although the information included in the reports might surely be used by various governmental agencies when formulating regulations, mere inclusion in the report in no way changes a chemical's legal status. Yet the chemical industry, fearing the potential for regulations that might cut into profits, has spent a small fortune attempting to limit the amount of basic information the government may distribute. The industry should be free to fight any regulations that it deems unjust, but it should pick an appropriate time to make its stand. With upwards of 400,000 Americans dying annually from cancer, and with evidence mounting that between one-third and two-thirds of all cancers are environmentally induced, it is far too self-serving and utterly inappropriate for the chemical industry to attempt to limit the spread of basic information.

The abuse of science can sometimes be incredibly blatant: in this particular case, the chemical industry is fighting to keep the federal government from sharing information that would assist scientists and policymakers in using their limited energies and their limited resources wisely. The situation is, however, not always quite this clear. Although, as the attacks on the Annual Reports on Carcinogens have shown, there are times when data are purposefully politicized to advance a particular end, there certainly are occasions when we find it virtually impossible to divorce politics, ethics, or a myriad of other issues from the data. A striking, if deeply troubling, example of just this conundrum concerns the uses to which data acquired by Nazi scientists under barbaric circumstances should be put.

The situation, so simple that it could come from a basic textbook, is an ethicist's nightmare. Years ago, truly terrible experiments were conducted on human beings. Data were obtained by immoral and despicable means. Today, however, those data, if used correctly, have the possibility of saving lives. Should the data be used? Or would their use glorify the atrocities that were committed and encourage people in the future to undertake similarly heinous experiments? The scenario outlined, and its associated ethical problems, are quite real. After much soul searching, the U.S. Environmental Protection Agency (EPA) has decided not to use the data.

The information in question comes from Nazi experiments conducted on inmates in concentration camps. Between 1943 and 1944 at Fort Ney near Strasbourg, France, Nazi doctors exposed prisoners to the toxic phosgene gas in ghastly experiments designed to search for a possible antidote. Phosgene, which had been used as a chemical weapon during the First World War, causes irritation and fluid accumulation in the lungs, making breathing difficult, if not impossible. Today phosgene is widely used in the manufacturing processes of plastics and pesticides, and according to *Science* magazine, about one billion pounds is produced annually in the United States. The EPA is concerned about the health risks faced both by plant workers and by people living nearby. ICF-Clement, Inc., an environmental consulting firm, made use of the Nazi data while undertaking a risk assessment of the chemical for the EPA. After twenty-two EPA employees signed a letter protesting the use of the data, EPA administrator Lee M. Thomas barred the use of any Nazi data acquired from human experimentation. "To use such data debases us all as a society, gives such experiments legitimacy, and implicitly encourages others, perhaps in less exacting societies, to perform unethical human 'experiments,'" wrote the critics. They also claimed that the experiments were so poorly done that the data were essentially worthless.

A number of people at the EPA and at the consulting firm disagree strongly. Todd Thorslund, a vice president for ICF-Clement, said, "Of course nobody in their right mind condones the experiment. The question is, given that this fiendish thing was done, what do you do with the information that exists? I suspect that the prisoners would have wanted to have the information used to help somebody." The Nazi data are the only available experimental information concerning the effect of phosgene on humans. Although the "experiments" were poorly designed, Thorslund and Ila Cote, an EPA toxicologist who was partly responsible for the study, have pointed out that if similar information came from an industrial accident exposing workers to the chemical, it would be invaluable. Because the human data have been ruled off-limits, determination of the risks associated with low levels of phosgene will await animal studies. As with most animal experimentation, the applicability to humans remains an unanswered question.

It is hard to fault the EPA for its position against using the Nazi data; nothing could be worse than to legitimize, in any way, any such atrocities. At the same time, the data are available, and they do provide information that cannot be acquired in any other fashion. That information might help determine what, if any, are safe exposure levels to phosgene. Is it fair to those people currently being exposed to the chemical to pretend that applicable data do not exist? Can the ethical questions be so compelling that we ignore information that might conceivably reduce the amount of human suffering and misery currently being experienced? As I said, it is an ethicist's nightmare. On balance, and without minimizing the heartfelt concerns of critics or the suffering of the inmates, I agree with Dr. Cote: "My personal opinion is that when data is collected in an unethical fashion, if it is important in protecting public health and is not available in any other way, I would use it."

Although the situations are so dramatically different, the central principles underlying the use of the Nazi data and the publication of the Annual Reports on Carcinogens are the same. The Nazi data, independent of the actual experiments, are value neutral, and thus if there is a productive use to which the data can be put, they should be put to that use. Similarly, the value-neutral data in the Annual Reports on Carcinogens, independent of the chemical industry's fears, can be put to productive use. The important point is that nothing is gained when available knowledge is censored.

This does not mean, however, that knowledge should be acquired at any cost. In this light, look at the chemical industry's plans to perform indoor air pollution tests on human subjects. The situation revolves around the chemical chlordane, a highly toxic insecticide. There is not much controversy over the effects that chlordane can have on humans. Most everyone agrees that it is a human carcinogen causing, among other problems, leukemia, aplastic anemia, convulsions, miscarriages, and birth defects. In 1978 the EPA took an almost unprecedented step and banned all agricultural uses, although it allowed the continued use of the chemical in homes for termite control. The logic was that when used to exterminate termites, chlordane should not come into contact with humans.

Years earlier concern had already surfaced that chlordane was escaping from building foundations and polluting the air inside treated structures. In the mid-1970s, for example, the air force found chlordane in the air of 80 percent of the buildings sampled. In 1979 a National Academy of Sciences toxicology panel undertook a review and concluded that it could not "determine a level of exposure to any of the termiticides below which there would be no biological effect under conditions of prolonged exposure." The panel recommended evacuating homes when levels of chlordane reached the point at which occupants would inhale a total of 5 micrograms. In 1982 the air force did just that, evacuating two hundred homes.

In 1983 a New York survey found high chlordane levels in 6.4 percent of the homes sampled, and the New York Health Department estimated that those levels would lead to between 1.6 and 5 additional cancers per 1,000 people. Samuel Epstein, professor of occupational and environmental medicine at the University of Illinois Medical Center, went even further. He called those figures "gross underestimates" and claimed that as many as 2 additional cancers per 100 people might be expected. In 1987, Diane Baxter, toxicologist at the National Coalition Against the Misuse of Pesticides (NCAMP), said, "If EPA doesn't have enough information to ban this chemical, then they'll never have enough information to ban any chemical."

Finally, in August 1987, the EPA entered into an agreement with Velsicol Chemical Corporation, the inventor and sole manufacturer of chlordane. The agreement stipulated that Velsicol would stop all sales on 15 April 1988 and make no further sales until it could demonstrate that it could develop a delivery system that would not release any chlordane into the air within the treated building. In mid-June 1988, Velsicol, with EPA approval, was ready to proceed with its tests. The absurdity was that Velsicol was planning to conduct its tests on two hundred inhabited homes around the country. NCAMP's director, Jay Feldman, accurately stated that this amounted to "EPA-sanctioned human testing of a pesticide." The company adamantly, if irrationally, refused to concede this point. Donna Jennings, a company spokesperson, repeatedly stressed to me that the company believes that chlordane is in no way dangerous to humans and that, therefore, humans are not being experimented upon. Its goal, she argued, is to demonstrate that chlordane is perfectly safe.

Quite simply, Velsicol's position was morally and ethically indefensible, and the EPA's failure to prohibit the tests from proceeding in this manner is a disgraceful abnegation of responsibility. There was absolutely no need for the test to be performed in inhabited houses. If the tests prove unsuccessful and residential atmospheres are contaminated, chlordane will not be allowed to be used in the future, but what would the government say to the families involved in the experiments? What sort of nonsensical logic has our government employed in sanctioning these experiments? And what sort of morality allows a company to propose such experiments?

(As an aside, it is worth noting that, to some extent, the focus on Velsicol's unconscionable experimental design has distracted the public from a more important issue. Chlordane, although banned from agricultural use in 1978, is still regularly found on potatoes, and serious contaminations of cattle feed as well as both cow's milk and human mother's milk have recently been reported. Fears abound that the insecticide, which has been used in about 700,000 homes annually, is making its way into the water supply. Obviously the EPA has been ineffective in its control of the dissemination of this toxin.)

Simply put, the experimental design calling for the newly designed delivery system to be tested in inhabited homes was abhorrent. Until the system can be conclusively shown to be safe, there is no justification for a testing protocol that calls for humans to serve as guinea pigs. The issue is purposefully complicated in this case because the chemical company was claiming to be undertaking scientific experiments—as if such a claim completely absolved it of any responsibility for atrocious experimental design. All of that is bad enough, but what is really frustrating is the type of response elicited by a column I wrote summarizing exactly these points and calling for a more rational and more humane experimental design.

Editor Jerry Mix of *Pest Control*, a mouthpiece for the pesticide industry, responded to that column with an editorial entitled "The Zimmermans have to be told which way is up" (Sept. 1988). In that diatribe, he completely ignored my charge that the experimental design chosen by Velsicol was unnecessary and unethical, as well as the fact that Velsicol itself had signed an agreement with the EPA not to use chlordane until it could demonstrate a delivery system that kept the chemical out of the air in homes. Instead, Mix decided to charge that chlordane was actually safe and that I was intentionally bending the truth about its health effects. Instead of dealing with the scientific issue at hand, Mix urged exterminators to "take some time and combat people like Zimmerman."

His readers responded to his exhortations by writing letters to me, like the one from a registered entomologist and exterminator in Athens, Georgia: "It is people like you who do the general public the disservice of causing higher prices and a poorer product because they pass on through scare tactics misinformation." Or like the one from an exterminator with a Ph.D. living in Birmingham, Alabama: "You are supposed to be a smart man; you have the credentials. Apparently, you have been educated beyond your intelligence."

In spite of such anti-intellectual responses to a criticism of scientific methodology, some good was accomplished. Because of a large amount of extremely negative publicity (a very small portion of which might have been a result of my column), Velsicol was unable to find a contractor willing to undertake the experiments desired. The company then backed away from its position and shelved the idea of reintroducing chlordane into the housing market.

One of the most troubling aspects of the turn to anecdote, superstition, and politics in place of scientific knowledge is that the turn is most likely to occur in a crisis situation. In many ways, it is perfectly understandable that desperately ill people will grasp at any straw, however unlikely it might be that that straw will provide meaningful improvement. Similarly, then, it should be equally understandable when society as a whole grasps at the same type of straws in time of crisis. The massive drought experienced by large portions of the United States during the summer of 1988 was one such crisis, and the response, from private citizens as well as from far too many public officials, was to turn completely away from scientific explanation and understanding toward a hodgepodge of makeshift spirituality.

As the drought in the Midwest deepened that summer and the ground dried out a little more, and as the stifling lack of moisture stunted the soybean and corn plants, our ability to reason seemed to wilt as well. Events in Clyde, Ohio, were typical of the region. Regular prayer meetings begged for rain, and Leonard Crow Dog, a Sioux medicine man, along with a number of associates, was brought in from South Dakota to conduct traditional rain-seeking rituals. Crow Dog guaranteed rain, along with lightning and thunder, within the week. In nearby Millersville, the Reverend Omer Rethinger prayed for rain a number of times and asked, in a sermon, "Do we really believe God can make it rain, or do we depend too much on modern sciences for answers to our current dilemma?" Governmental officials offered similar counsel. Although then Secretary of Agriculture Richard E. Lying may not have been completely serious when he told the Senate Agricultural Committee, "I guess the best thing for us to do is pray for rain," it is all too clear that Representative Dwight Wise Jr. (D-Ohio) was when he said, "The prayer of people, including the Indians, will help."

Prayer and ritual might be important personal responses to situations beyond our control, but it is all too clear that those individuals calling for prayers and dances did so not for spiritual release but for rain itself. Such calls are not appropriate public responses, for they place superstition and religion on a par with science in an attempt to understand the natural world.

Although, at the time prayers were offered, the drought was devastating, we should not have squandered the opportunity it provided to teach us about the physical laws of this world. By understanding that the drought in the Midwest was caused by a Bermuda high-pressure cell that was stalled farther east than usual over the eastern seaboard and that the cell deflected moisture-laden air from the Gulf of Mexico farther west than usual, we would have made ourselves more able to grasp the large-scale nature of weather patterns. From there it would have required only a small step to recognize that environmentally we really do live in a global community.

Instead, too many of us look toward the supernatural when faced with phenomena that we do not understand. Prayer and ritual dance may do wonders for our spirit, but they will teach us nothing, and they certainly will not help the wheat and sorghum. Is it any wonder that our children fare so poorly on standardized science exams relative to students in most other industrialized nations when we are so quick to turn our backs on understanding?

Although neither science nor scientists can bring an end to a severe drought, that does not mean that they are useless. Science, if we permit it, may provide the understanding and the knowledge that will enable us to predict future droughts and to allow us to better plan for and cope with severe water shortages.

Mine is not an antireligious message, for I recognize the enormous good that religion might do in the lives of many people. My message, rather, is a plea for rationality even when it comes to our needs and desires. We cannot forsake meteorology for rain dances any more than we can give up medicine for faith healing.

Not surprisingly, the high-pressure system causing the 1988 drought moved eastward out to sea, and rain fell again. Many were quick to credit their rain dances, their prayers, and their incantations for our good fortune. As the rains replenished our mighty rivers, water levels again rose. When, less than two years later, an overabundance of rain caused a tragic flash flood that devastated the town of Shadyside, Ohio, killing 26 people, none of the "rainmakers" were willing to take credit for that.

I recognize that science alone is not going to solve our environmental or our technological problems. Obviously, politics, current events, and incidental crises all will affect the way we choose to take responsibility for the health of ourselves and of our planet. But I further recognize that those problems are not going to be solved at all if we ignore scientific methodology and the knowledge that it has brought us. We turn toward pseudoscience, superstition, and political expediency at great peril to ourselves and our environment.

Chapter Four

ॐ

Political
Science

The Role of
Government in
the Scientific
Process

In late December 1988 a member of the National Science Foundation's (NSF) Directorate for Science and Engineering invited me to give a talk at the agency's Washington headquarters on the creationist threat to science education. I was delighted, both because I was going to be in Washington to serve on an NSF panel reviewing grant applications and because my research on public perceptions of the evolution-creation controversy had turned up statistics that would send shivers through anyone dedicated to public science education.

For example, research I had published just the preceding year found that 15 percent of the high school biology courses in Ohio presented the basics of creationism in a favorable light and 40 percent of the biology teachers polled indicated that they favored the teaching of creationism in public schools. My study went on to ask teachers whether they had experienced any pressure to alter the course content of their science classes. Pressure was reported to have come from a variety of sources: religious figures, school administrators, parents, and even God. I found that creationists applied significantly more pressure to teachers than did proponents of evolutionary theory. Although the impact of creationism is likely greatest in the rural south, I discovered that less time was devoted to teaching evolution in high school biology courses in the 1980s in this northeastern, industrial state than was devoted nationwide in the 1940s. Perhaps even more striking, my research found that many of the managing editors of the nation's daily newspapers did not know that humans appeared on earth millions of years after dinosaurs had died out, or that the earth is billions, not thousands, of years old, or that teaching evolution is a scientific rather than a religious endeavor. Yes, I was very pleased to have been offered the invitation to speak to leaders at the National Science Foundation.

You can imagine my shock, then, when I arrived on 4 January for my presentation and was told that, unlike other talks scheduled at NSF headquarters, mine had not been permitted to be advertised publicly because the topic was considered too controversial and possibly offensive to members of Congress and their staffs. Typically, at the NSF and at colleges and universities around the country, posters announcing such a talk would be hung in prominent places. No such posters were permitted for my talk; instead, advertising was limited to word of mouth. At first, I found the thought of the National Science Foundation's downplaying a talk on a topic so central to the nature of science education to be incredibly appalling. I likened it to the Environmental Protection Agency hiding a talk on the dangers associated with global warming or the Food and Drug Administration covering up a talk on the possible relationship between birth control pills and the development of breast cancer.

Upon further reflection, however, I realized that my anger was misdirected. Rather than indict the National Science Foundation for declining to publicize the fact that the controversy and ignorance swirling around the evolution-creation issue are seriously compromising science education, I should blame the political nature of the process leading to the funding of science education initiatives. A public notice of such a talk would inevitably have come to the attention of numerous misinformed congressional aides and members of Congress who regularly enter NSF headquarters, and the bias of some might affect their subsequent budget decisions.

This is no idle fear when you consider the results of a congressional survey I conducted. I mailed legislators a questionnaire with thirty declarative sentences and, using standard social science methodology, asked respondents to check one of five options for each statement: strongly agree; mildly agree; no opinion; mildly disagree; and strongly disagree. My results indicated that approximately 53 percent of the members of Congress responding to the questionnaire did not disagree strongly with the statement "Creation science should be impartially taught in public schools" and that slightly more than 72 percent of the respondents did not disagree strongly with the statement "Some people can accurately predict future events with psychic power." And, I should hasten to point out, my questionnaires were sent out *after* all the negative publicity concerning the astrological beliefs of first lady Nancy Reagan.

Even more to the point, the recent past shows us what can happen to support for science education when those unable to distinguish between science and pseudoscience gain the political upper hand. In 1981 and 1982, during the first two years of the presidency of a man who publicly confessed his skepticism of evolution, the budget for the NSF's Directorate for Science and Engineering Education was reduced from $81 million to virtually zero. (The only reason the budget didn't go all the way to zero was that some of the money remaining uncut had already been committed for student fellowships and no one could figure out a politically acceptable way to recall those funds.) The decision to cut out the education component of the NSF, despite the fact that the agency was originally established in 1950 to strengthen the nation's commitment to both basic science and science education, was made at the Office of Management and Budget with no input from the National Science Board or the director of the National Science Foundation.

The impetus for such massive cuts is directly attributable to the foundation's support through the early 1970s for development of a highly successful curriculum for fifth and sixth graders entitled "Man: A Course of Study" (MACOS). The year-long program was designed to teach students to think scientifically within an evolutionary context. Although the program was acclaimed by teachers and was adopted by about seventeen hundred schools nationwide, it fell afoul of religious fundamentalists who objected to its emphasis on evolution. They ran such a slick grass-roots media campaign that the NSF was forced to withdraw MACOS. Consider the claptrap that syndicated columnist James Kilpatrick wrote about MACOS in April 1975, when he called it "an ominous echo of the Soviet Union's promulgation of official scientific theory" (*Boston Globe*, 2 Apr. 1975).

It was utterly ridiculous to equate the NSF's efforts to improve the quality of scientific reasoning in elementary schools with T. D. Lysenko's thirty-year refusal to permit any research on evolutionary biology and Mendelian genetics in the Soviet Union because he felt that the subjects had too many capitalistic overtones. It was particularly disingenuous of Kilpatrick to do so because one of the biggest complaints that fundamentalist critics leveled against MACOS was that its curriculum was far too open ended—simply, it encouraged students to ask too many questions. Yet in 1981, when the ax fell on the Directorate for Science and Engineering Education, members of Congress used code words indicating that they were in sympathy with Kilpatrick's outrageous position. The cuts were being made, their arguments went, not because support for science was a bad thing but rather because it was important for the scientific content of curricula to be determined at the local level—as if human beings might have evolved differently in Peoria than they have in New York.

Given this record, NSF officials were wise to be wary of congressional opinions, regardless of how ill conceived they might be. My research indicates that, from the perspective of science education, the problem is far worse than it appears in Washington because the level of scientific sophistication at the state level is lower than at the national level. On a range of questions dealing with pseudoscientific topics, federal legislators, as ignorant as they were, provided less extreme responses than did their colleagues at the state level in Ohio. Consider the following six results:

— only 45 percent of the Ohio lawmakers and 65 percent of their federal counterparts strongly disagreed that "aliens from other worlds are responsible for the construction of some ancient monuments";

— when presented with the statement "It is possible to communicate with the dead," 51 percent of the federal officials and 35 percent of the Ohio officials strongly disagreed;

— whereas 74 percent of those serving on Capitol Hill strongly disagreed that "astrology is an accurate predictor of future events," only 48 percent of those serving in Columbus made such a claim;

— 56 percent of the federal legislators disagreed strongly with the statement "Dinosaurs and humans lived contemporaneously," but only 22 percent of the Ohio legislators felt similarly;

— approximately 53 percent of members of Congress and 81 percent of the Ohio officials did not disagree strongly with the statement "Creation science should be impartially taught in public schools";

— when given the opportunity to say whether they "accept the premises of creation science," 60 percent of the federal lawmakers, but only 19 percent of the locals, strongly demurred.

Results of this sort go well beyond individuals forgetting some arcane facts and speak wonders about the twin roles of science and education in a democratic society. Is it surprising that local legislators having such a poor conception of science are often loath to fund schools at the levels necessary for meaningful learning to occur? Aside from the fact that members of the House and Senate are our leaders and that leaders, as role models, should be our ultimate teachers, our federal government really does not have much direct impact on the specifics of science education. But the situation is far different at the local level, the level at which the decisions regarding curriculum, staffing, and classroom funding are made. The creationists, who are well aware of this, increasingly direct their message and calls for change at the most local of levels, the school board, where grass-roots support by scientific illiterates is greatest. Again, research of mine presents a chilling story. Approximately 53 percent of the Ohio school board presidents that I surveyed in 1986 stated that creationism should be favorably taught in the public schools, but only two-thirds thought that evolution should be favorably taught. Slightly less than half of the respondents indicated that they personally accepted the theory of evolution, the single most important unifying concept in all of biology. Whenever the creationists win a battle, more schoolchildren lose the critical opportunity to learn to think creatively and effectively. Unfortunately, many such battles are being waged daily.

At the national level, the issue is slightly more abstract but no less important than at the local level. The major question at the federal level, as was so clearly evident in the MACOS situation, is how limited federal research dollars are to be spent. Legislators seem somewhat schizophrenic on this issue. They seem incapable of deciding whether scientific experts or elected members of Congress should be determining the distribution of research awards. The debate has gotten so out of control that serious damage is being done to the nation's ability to fund critically needed research initiatives.

In pressing for a line-item veto in his final State of the Union address, President Reagan criticized Congress for appropriating money for various scientific projects. "For example, there's millions for items such as cranberry research, blueberry research, the study of crawfish and the commercialization of wildflowers," he chastised. Although it is difficult to take seriously scientific advice from the man who claimed that trees were responsible for acid rain, and although many scientists stepped forward to defend these expenditures, the president was not too far from the mark. Though he was entirely wrong to criticize these projects for what he believed to be the frivolous nature of cranberries, blueberries, and crawfish research, he was correct to criticize the funding. His complaint, however, should have centered on the process by which Congress funds scientific research rather than on the specifics of the projects chosen for funding.

Money for these projects was part of a nationwide package of scientific support totaling about $218 million which Congress approved in the waning days of 1987. In 1992, a whopping $105 million was appended at the last minute just to the defense appropriations bill for scientific projects favored by members of Congress. In 1993, that figure rose to $178 million. In 1991, Representative George Brown (D-Calif.), chair of the House Science, Space, and Technology Committee, admonished his colleagues for the actions they took in crafting the NASA portion of the federal budget. The final bill, he claimed, contained "over $100 million in projects that were never requested by the administration, never authorized, and never discussed on the floor." In no uncertain terms, Representative Brown concluded, "I believe this practice has simply gone too far." Representative Brown was absolutely correct. Such large commitments of money, even coming at a time when basic scientific research in this country is woefully underfunded, is not welcome news.

The problem is quite simple. The funding decisions were made on political rather than scientific grounds. Congress, year in and year out, unilaterally dictates that various governmental agencies (in 1987, the year to which President Reagan referred, the agencies included the departments of Agriculture, Energy, Commerce, Interior, and Transportation, the Environmental Protection Agency, the Federal Aviation Administration, and the Bureau of Mines, among others) fund specific projects even though those agencies had not requested the money to do so. Furthermore, the projects selected did not undergo the normal process of review by scientific experts.

The general procedure for allocating federal money for scientific research is quite straightforward. Under normal circumstances, an institution or a scientist, in response to a well-advertised solicitation, submits a formal proposal to a federal agency. Generally, the proposal is then sent to a good number of experts who offer their professional opinion on its scientific merit and applicability as well as on the credibility, talents, and record of the applicant. After reading and evaluating all of the proposals as well as all of the reviewers' comments, a panel of additional experts convened by the granting agency ranks the proposals in terms of their scientific merit and importance. Projects are funded in decreasing order of merit until the money runs out. Unfunded proposals are returned along with comments from the expert reviewers. Those comments often help to improve the quality of the science itself and thus make the proposals more competitive next time around.

When Congress appropriates money for specific projects, it neither seeks nor receives the benefit of expert scientific advice. In the absence of such advice, political considerations rather than scientific ones become central to funding decisions. How have we come to this absurd point where pork barrel politics plays such a role in dispersing limited research funds? By allowing narrow interests to outweigh those of the nation as a whole.

Lawmakers self-servingly recognize that the economic health of their districts can be linked to investments in science and education and that fighting for projects located in their districts will be in their best political interest. Leaders of academic institutions, as well, perceive direct lobbying to be in the interests of their institutions. With very tight federal funding of research, only a small percentage of meritorious proposals will be funded at all. The odds of receiving federal dollars by lobbying Congress thus often seem better, or at least no worse, than those associated with the competitive routes established by agencies such as the National Science Foundation and the National Institutes of Health (NIH). The scientific community, and therefore society in general, however, are much worse off when many of the limited dollars available for science are spent in this manner.

The peer review system for scientific proposals is by no means perfect. Scientific bandwagons, like political bandwagons, occasionally pass through the scientific community, and certain unworthy but faddish proposals are funded. Such mistakes are insignificant, however, when compared with the potential for misuse when funding decisions are purely political. Who can forget the spectacle of members of the House Science, Space, and Technology Committee falling all over themselves trying to be photographed with B. Stanley Pons and Martin

Fleischmann in the days following their announcement of the discovery of cold fusion? Publicity of this sort is fine, but the lawmakers also repeatedly asked the chemists what sort of support they needed to continue their work. Pons and Fleischmann, well prepared for this sort of attention, brought along Chase N. Peterson, president of the University of Utah, who politely requested a mere $25 million to keep the experiments going. Although the politicians were entranced, most scientists did not believe that the cold fusion pronouncements would stand up to peer review—and indeed they did not.

The consequences of political involvement in the scientific process can be greater than federal officials leading an ill-fated parade or unwisely spending a few million dollars here and there. As we have seen with the federal ban on fetal tissue research, when political considerations become paramount, free inquiry and the public welfare often take a back seat. Daniel E. Koshland Jr., in an editorial in *Science,* the journal he edits, addressed the interface between politics and science directly:

> Fetal tissue research should have as little relation to the abortion issue as organ transplants have to automobile accidents or suicides. Tissues and organs from automobile accident victims played a major role in the development of sound organ transplant therapies, but no one argued that such research justified or legitimized automobile accidents. In the same way, the decision to use fetal tissue for research is made separately from the decision of an individual to have or not to have an abortion. No scientist should be involved in soliciting an abortion in order to obtain material for research, but to limit materials for research on the grounds that one is taking a stand on the abortion issue is an illogical and counterproductive path that can only damage all participants. (26 June 1992, 1741)

Human fetal tissue is a unique resource. Because it is still undergoing differentiation, it has an enormous amount to teach us about the process of development. Additionally, because of this trait, it is much more adaptable to diverse experimental conditions than is adult tissue. All of these factors have led fetal tissue to play an important role a wide range of medical research programs from polio to AIDS. Researchers are also beginning to find therapeutic uses for fetal tissue in the treatment of some maladies such as Parkinson's disease.

Similar devastations have occurred in women's health care and AIDS research, to name just two other scientific issues that got caught up in the political process. The large bulk of the medical research conducted at or under the auspices of the National Institutes of Health ignored women until a number of members of Congress and NIH director Bernadette Healy began to make a fuss. And as Randy Shilts so dramatically points out in his book *And the Band Played On,* federal funding of AIDS research was almost completely held hostage to the political process for years. Funds began to be released more freely only when it became clear that AIDS was not exclusively a "gay disease." When politicians rely on politics rather than on scientific peer review, we find ourselves with institutional neglect of vital areas of medical and humanitarian research.

A criticism leveled against the peer review process is that when scientific proposals are judged solely on their merits, an even geographic distribution of resources is usually not achieved—some states, indeed some whole regions, remain underfunded. There can be no denying that this is the case, but we need to recognize that the funding of scientific initiatives in our country should not be akin to a federal welfare program. The best science should be supported independent of where in the country it is being conducted.

Shortly after his retirement from the Senate, William Proxmire (D-Wis.), long a vocal critic of pork barrel funding of scientific research, said at a talk at Oberlin College that he was aware of only two ways to set priorities for funding: congressional pork barrel and scientific peer review. He concluded that the former was a disaster and that the latter needed strong backing if our worldwide lead in scientific endeavors was to be maintained. Ironically, some of his most public actions served to undercut dangerously the very system he wanted to preserve.

Senator Proxmire, you will remember, was the creator of the Golden Fleece award. Nearly every month for approximately fourteen years, Proxmire awarded a Golden Fleece to some agency of government for apparently wasting taxpayers' money. Certainly, quite a few of these awards were well deserved. The navy, for example, clearly earned its Golden Fleece for paying $792 for a designer doormat.

But not all of the senator's choices were of this sort. Occasionally he went after research scientists, particularly those in the behavioral and social sciences, whose projects, although they might, to the uninitiated, sound just as ridiculous as a $792 doormat, were dramatically different. Every research project that has won a Golden Fleece, from a

National Institute of Mental Health–funded study of Peruvian brothels to a study of the reproductive behavior of Japanese quail, was funded only after the granting agency decided that the particular project in question was among the very best submitted using the process of peer review outlined above.

Robert L. Park, executive director of the Washington office of the American Physical Society, has been quite critical of Proxmire. "Anything people don't understand can be made to look foolish. He took advantage of that repeatedly." What is so troubling is that pure research in almost every field, from animal behavior to nuclear physics, is so esoteric and apparently so removed from any practical application that, whatever its value, it might well appear to be a foolish waste of taxpayers' money. Yet scientific knowledge accumulates in small steps, by researchers who pose extraordinarily narrow questions. Consider, for example, that scientists rearing baby rats in isolation and simply stroking their backs have greatly enhanced our understanding of the processes of human physical and psychological development. Politicians do not have the expertise to evaluate the significance of the many small, but critical, scientific steps represented by most research proposals, and they should not pretend that they do.

Proxmire's attack on the social and behavioral sciences has caused great harm. Because of the Golden Fleece awards, certain types of research quickly fell out of favor among funding agencies. Alan G. Kraut, executive director for science at the American Psychological Association, notes that AIDS experts, now stressing education to alter high-risk sexual behavior, are at a disadvantage because research on attitudinal change has been ridiculed for so long by nonexperts like Proxmire. Additionally, Proxmire received so much favorable attention from his antics that Congress has, too frequently, attempted to step in and substitute its opinions for the wisdom of scientific peer reviewers. The most notorious escapade occurred on 9 April 1975 when the House of Representatives, by a vote of 212 to 199, adopted an amendment to an NSF appropriations bill. That amendment called for the direct involvement of Congress in the review process of all NSF grants and gave it veto power over all proposed awards. Not only are members of Congress educationally unprepared to undertake any sort of scientifically sophisticated review of the highly technical grant applications, they surely do not have the time to handle the approximately fifteen thousand proposals their vote called into question. Happily, a Senate conference committee deleted the amendment before it came to a vote of the full Senate.

When politicians play hardball, there are times when pork barrel and peer review funding come into direct conflict with each other. One such incident occurred in May 1992 when Senator Robert C. Byrd (D-W.Va.), chair of the Senate Appropriations Committee and a master at bringing pork home to his state, introduced a bill to block the funding of thirty-four research projects. All thirty-four of those projects had received very favorable peer reviews at either the NSF or the NIH and were to receive money from those agencies. What set Senator Byrd off was that a couple of months earlier President Bush had attempted to force Congress to cut $30 million of pork for sixty-seven unreviewed projects. That Byrd's bill, which was approved by the full Senate, was nothing more than political retaliation for Bush's action can be clearly seen from the comments of an Appropriations Committee staffer quoted in *Science:* "Two can play this game." Jack Lein, vice president of health sciences at the University of Washington, spoke for the entire scientific community when he lamented, "Right away this tells me that what we've held near and dear—the peer review system—could well go right down the tubes."

This is no time to be cutting support for basic scientific research in this country. Cuts made during the Reagan-Bush years have brought rates of funding for proposals submitted to the NSF and NIH to an all-time low, having plummeted steadily during those two presidential administrations. And, on average, even those grants that are awarded are buying less than they used to buy. When adjusted for inflation, the average grant award has decreased steadily over the past decade. The scarcity of federal research funds has led to a serious morale problem among young scientists, many of whom have chosen to drop out of the scientific establishment altogether. Society stands to lose much if these individuals, people who have just earned their Ph.D.'s, who should be entering the most productive years of their professional lives, opt for different careers. David Blake, senior associate dean at the Johns Hopkins University School of Medicine, stated his concern quite bluntly: "We're going to turn off an entire generation of new biomedical scientists just at a time when we need them."

Another area in which politicians and scientists seem to be in conflict is the issue of scientific fraud. When a high-level congressional committee headed by Representative John Dingell (D-Mich.) holds numerous public hearings about scientific fraud and when the Secret Service is brought in to examine laboratory notebooks of research scientists, the public has a right to assume that lies are rampant in the scientific community. Furthermore, since large amounts of the research conducted at colleges and universities are funded with public money, it is not unreasonable to demand that controls be put in place to eliminate such fraud. Much of the attention is, however, greatly misdirected, and, as is so often the case when politicians look for simple answers to complex problems, many of the proposed solutions will be detrimental to the scientific process.

Two points that need to be made about scientific fraud concern its frequency and its definition. Because fraud is exceedingly difficult to document, accurate estimates of its frequency are impossible to achieve. Scientists, not surprisingly, tend to think that it is rare, whereas journalists and some members of Congress, perhaps along with a growing percentage of the general public, think that it is far more common. Dr. Marcel C. LaFollette, in her comprehensive 1992 book *Stealing into Print: Fraud, Plagiarism, and Misconduct in Scientific Publishing,* says that she cannot give an accurate assessment of its frequency. Nonetheless, I believe that it is probably not as common as some politicians would have us believe. Nor, however, was it as uncommon as tradition suggests. Charles Babbage, designer of a prototype calculator, was concerned with the subject of fraud as far back as 1830 when he wrote a piece entitled "Reflections on the decline of science in England, and on some of its causes," in which he complained about scientists who make "improvements" in their studies by "cooking" their data. There is a very strong suspicion, for example, that Gregor Mendel, the Austrian monk of a century ago who was the founder of the science of genetics, "massaged" much of his original data to such an extent that he crossed over the line of scientific propriety. Similarly, there can now be no doubt that Sir Cyril Burt, an early leader in the debate over whether genetics or environment primarily determines intelligence, fabricated huge amounts of data.

Yet I believe that outright scientific fraud is and has been relatively rare. The reason is quite simple: in science, it is extremely difficult to get away with fraud. The prerequisites for acceptance of any scientific idea are repeatability and independent verification. It is clear, for example, that the existence of cold fusion was called into question almost immediately because no investigators were able to independently verify the results originally reported at the University of Utah. The cold fusion example is not one of fraud but rather of investigators finding in their data what they desperately wanted to believe whether it was there or not—and then orchestrating a public relations blitz to announce their findings. No fraudulent experiment will ever be able to long stand up to the sort of testing which is always directed at exciting new scientific results. How common can outright fraud be when the long-term costs of misrepresentation so clearly outweigh any possible short-term gains?

Part of the confusion may arise from the definition of fraud. Scientists are surely no less human than members of any other profession. They are subject to the same type of pressures about career advancement, higher pay, and public recognition as are others. Because of these pressures, misconduct occasionally (frequently?) occurs. Make no mistake about it. Unfortunately, academic science can be every bit as political as the political arena. Scientists have been known to appropriate the work of others without due recognition. They have been known to build their own reputation at the expense of others. They have been known to ignore previous work in an attempt to claim precedence in an area. None of this is moral, acceptable, or forgivable. But neither is any of it worthy of full-scale congressional investigation. All are examples of professional misconduct and should be dealt with appropriately, but none, in my mind, are examples of fraud with a capital *F*. None call into question the very nature of the scientific enterprise.

Another two definitional problems also stem from the fact that scientists are human. First, like all humans, they make errors—errors of commission, of omission, and of interpretation. Neither sloppy science nor honest error should be considered fraudulent, however. Second, scientists are subject to self-delusion like everyone else. Sometimes they honestly see what they want to see, whether it is there or not. As discussed in the previous chapter, good scientific methodology calls for double-blind experiments whenever possible just to minimize this problem. However, not all experiments can be designed in this fashion, and we should not overreact when scientists read too much import into any one of their experiments.

It is worth pointing out that scientists, by their very nature, are a skeptical and cantankerous lot; they doubt much of what they read and virtually all of what they hear. Arguments and disagreements over data interpretation and experimental methodology abound in the scientific literature. This sort of controversy is perfectly healthy and is what makes science so vibrant, powerful, and fun to be a part of. But such public disagreement is a far cry from fraud.

Now that Congress has gotten involved, solutions to the "fraud problem" have been advanced. One has focused on the process of peer review of scientific papers. Like peer review of grant proposals, peer review of publication requires that a journal editor send each submitted manuscript to a number of experts for evaluation. Each expert will read the paper, assess its significance, methodology, and logic, and make a recommendation about publication. The suggestion has been advanced that reviewers undertake audits of the original data used in the paper as well as replicate the experiments prior to publication. The makers of such a proposal exhibit gross naivete about the review process. A favorable review of a manuscript does not mean that the ideas and hypotheses included in the paper are or will be correct; rather, it means that the ideas are well enough formulated to warrant further, detailed scrutiny by the scientific community at large. Even more important, if such a suggestion were put into practice, it would bring science to a standstill.

I currently receive an average of one manuscript to review per month. Each review, for which I receive absolutely no compensation of any sort (other than feeling an active participant in the scientific community), takes, on average, a full day—a full day that I cannot use to conduct my own research or interact with students and colleagues. The thought of repeating the experiments, some of which took years to complete in the first place, in each manuscript that I review is so absurd that it can be dismissed without comment. Even to examine the voluminous raw data that went into each paper, however, would be impossible. No editor would dream of making such a request of a reviewer.

None of this is to say that the scientific community has handled charges of fraud particularly well in the past. Clearly there are instances when some of our leading research institutions have less than forthrightly investigated charges leveled against scientists on staff. But institutional procedures for dealing with charges of scientific misconduct have been greatly improved of late, largely because of public outcry and concern. The nation will not be well served by the dramatic restructuring of a scientific community that a few have found guilty without trial and with scant evidence. Congress needs to be very careful where it treads.

Yet there clearly is an important role for governmental officials to play in the scientific process. It is up to our elected officials to look at the big picture and set the direction for our national research efforts. Rather than worry about individual research projects, our leaders and representatives should be thinking about research initiatives that will best serve the country. This is a much more difficult task than finding fault with a study of Peruvian brothels but one that is too important to be left to the scientists alone. We need real leadership if we are to make meaningful progress on such critical topics as issues of women's health care, AIDS, and environmental protection and restoration.

One such environmental direction that would be extremely productive for politicians to pursue is to make the Environmental Protection Agency an active rather than a reactive organization. Quite a number of people have unsuccessfully attempted to push Congress in this direction. In 1989 Erich Bretthauer, the acting chief of research and development at the EPA, wrote, "We must develop the capabilities to anticipate and prevent pollution, rather than simply controlling and cleaning it up after it has been generated." His remarks were part of a report entitled "Protecting the Environment: A Research Strategy for the 1990s," which took a far-reaching look at the EPA's involvement in the government's environmental science effort. In the terms of the report, the agency should move beyond a mission solely focused on an "end-of-the-pipe" cleanup.

Bretthauer recommended that new funding opportunities for basic environmental science research be established. Not only should more money be spent on such projects, but instead of all of the work being done in-house by EPA staff, an ongoing competition among nonagency scientists in academia and industry should be encouraged, much in the same way the National Science Foundation and the National Institutes of Health disperse federal research dollars. Such a program would broaden the types of environmental questions currently being asked and would increase the likelihood that significant advances would be made in new and innovative areas. Four major funding areas were targeted: human health risks, ecological risks, risk reduction methods, and basic research grants.

A look at a list of the nation's major environmental and health problems demonstrates the wisdom of the Bretthauer proposal. From depletion of the ozone layer, to the accumulation of gases leading to global warming, to massive acidification of lakes and forest ecosystems, to toxic waste disposal, we are spending exorbitant sums of money in an attempt to rectify environmental problems that could perhaps have been foreseen and averted. And even those sums may not be enough. Additionally, in some cases the damage done may well be irreversible. Allocating money for research to prevent the next generation of massive environmental problems is obviously going to be money well spent.

Whether the Bretthauer proposal is enacted or something akin to a proposal coming out of the Clinton administration calling for the establishment of a freestanding agency to coordinate environmental research makes no difference; some movement in this direction is essential. Here is an area in which Congress can exert scientific leadership that will make a huge difference to the lives of many Americans.

Chapter Five

༄

Global
Environmental
Problems

A Small Look
at the Big Picture

We have reached the point, in terms of numbers and technology, where, whether we mean to or not, we are able to alter this planet in ways previously unimagined. The United States' National Academy of Sciences and the Royal Society of London, perhaps the two foremost honorary scientific societies in the world, recently came together to issue a sobering alarm: "If current predictions of population growth prove accurate and patterns of human activity on the planet remain unchanged, science and technology may not be able to prevent either irreversible degradation of the environment or continued poverty for much of the world" (cited in *Science,* Mar. 1992, 1357).

This alert was followed by a document entitled "Warning to Humanity" signed by fifteen hundred scientists from around the globe, including ninety-nine Nobel Prize winners, a dozen national academies of science, the Pontifical Academy of Science, and the director general of UNESCO. That document, released in November 1992, said in no uncertain terms that "human beings and the natural world are on a collision course," which "may so alter the living world that it will be unable to sustain life in the manner that we know."

These dire warnings are based on two basic postulates. First, the world is a finite place, and all of its physical resources are in limited supply. While the earth's renewable resources are being stretched thin by population growth, its nonrenewable resources are simply being used up. Once these latter resources are gone, they are gone forever. Second, technologically we are now capable of instantaneously altering the environment of the entire globe where previously we were capable of devastating only pieces of ecosystems at a time.

Since the first Earth Day in 1970, we have seen a net gain of more than 2 billion human beings. To put this number in perspective, it is important to recognize that 2 billion was the total number of people in the world in the early 1930s. If conditions remain constant, predictions are that the world's current population of approximately 5.6 billion will double by the year 2050.

Although the "population problem" has certainly received significant public attention, especially since the publication of Paul Ehrlich's *The Population Bomb* in 1968, grave misunderstandings of the problem are still rampant. The most important misunderstanding centers on the difference between percentage rate of growth and absolute rate of growth. Many misguided optimists have stressed that population growth is slowing because the percentage rate of growth has slowed a bit. From a yearly increase of approximately 2.1 percent in 1961, the rate at which the world's population grew fell slowly to about 1.7 percent in 1979. It remained roughly constant at that point until 1989, when it crept back up to 1.8 percent. While 1.8 percent is surely lower than 2.1 percent, if you buy into this, you'll miss the very heart of the matter. Imagine that your spouse weighs 300 pounds and gains 10 percent of his or her body weight in a single year. Up to 330 pounds, your spouse decides to go on a diet, and at the end of the next year, he or she declares the diet a success with a weight gain of only 9.5 percent. When you have the uncomfortable realization that actual weight gain for year two was 31.35 pounds (or 1.35 pounds more than the pre-diet gain), I'm sure you'll change your position.

In absolute terms, more and more people are being added to the world each year. In 1961, approximately 64 million people were added to the world, in 1972 approximately 76 million were added, and in 1989 approximately 94 million people were added. In fact, the absolute rate of increase has climbed steadily upward every year since the end of the Second World War. The approximately 94 million people being added to the world's population each year as of 1989 translates into 258,000 each day, or 10,750 each hour, or almost 30 every second. Our finite world simply cannot continue to support this rate of growth.

Too often the population problem is viewed exclusively as one of the developing world. While it is true that the population is growing faster in developing countries than in the developed countries, it would be unwise and unproductive to view the problem as one exclusive to the Third World. Just looking at numbers of people ignores the important link between limited resources, pollution, and population number. Although the desperately poor of developing regions often cause significant environmental degradation, typically by deforestation in search of wood to use for fuel and by being forced to eke out an existence on marginal lands ill suited for agriculture, citizens of the developed world pose an even greater threat to the well-being of the planet as a whole. Paul and Anne Ehrlich make this point clearly in their 1991 book *Healing the Planet: Strategies of Resolving the Environmental Crisis:*

> The United States is the world's most overpopulated nation. It is the world's fourth largest nation in population, now numbering more than a quarter-billion people, and the average American consumes more of the Earth's riches than an average citizen of any of the other "big ten" nations with more than 100 million people: China, India, the Soviet Union, Indonesia, Brazil, Japan, Nigeria, Bangladesh, and Pakistan. Furthermore, the technologies used by Americans to support their opulent consumption are environmentally damaging and needlessly wasteful of energy. Because of this combination of a huge population, great affluence, and damaging technologies, the United States has the largest impact of any nation on Earth's fragile environment and limited resources. (9)

Garrett Hardin, in his 1993 book *Living within Limits,* referring to the work of agricultural geographer Georg Borgstrom, points out another basic misconception related to the "population problem." Hardin argues that too many people, even those most highly educated, have yet to focus on the fact that we are experiencing a human population explosion because, worldwide, there are still areas that appear to be underpopulated. Claiming that too many misguided books and articles have been written using the title "Standing Room Only," he calculates the amount of space needed if the human population were in fact crowded together in standing-room-only fashion. Assigning a rectangle 3 feet by 1 foot to each of the five billion people in the world as of 1989, he figures that all could be accommodated "on a mere 556 square miles, just 46 percent of the area of Rhode Island, our smallest state. A perfect square, 24 miles on a side, could accommodate the world's entire population, standing up. Alaska, with an SRO capacity of 5 trillion, could accommodate a thousand times the present world population" (121).

As absurd as calculations of this sort are, from one perspective reality is not all that different. Hardin, using 1980 census data, calculates the amount of space, at ground level, associated with each person living in Manhattan. Ignoring the large debate about the extent to which the census undercounts homeless people, Hardin figures that each Manhattan resident was allotted "a square of about 7 yards by 7 yards. And each resident had to share 'his' space with commuting office workers and visitors from out of town."

The real purpose of this exercise is to demonstrate the absurdity of simply comparing people and area. Borgstrom coined the term "ghost acres" for the land that is needed to support each human being on the planet. Although as of 1980 1.4 million people lived in Manhattan, none of those 1.4 million were fully supported by the land on Manhattan Island, let alone by the 7 square yards assigned to each resident. In fact, each Manhattanite required approximately 9.1 additional acres: 1.9 acres of cropland; 2.4 acres of pastureland; 2.6 acres of woodland; and 2.2 acres of "other" land (land for such things as factories, roads, recreation areas). Although, as Hardin demonstrates, these ghost acres go largely unseen and unthought of by most citizens, they cannot be ignored because they are absolutely essential if quality of life is to be maintained. So the next time you find yourself in a stretch of uninhabited country, count off 9 acres for yourself and another 9 for your companion and see how much is really left over.

In a world in which resources are in finite supply, humans face the same constraints as do all other species; the environment sets a limit on the number of individuals that can be supported, and when that limit (called carrying capacity) is exceeded for an appreciable length of time, quality of life begins to deteriorate and then population size begins to decrease, often precipitously. We are being foolishly naive and risking incredible worldwide trauma if we persist in believing that human technology can, without limit, raise the carrying capacity for the human species. That there are limits to growth is as basic a law of nature as is gravity.

Unfortunately for us all, basic economic theory largely ignores the basic biological reality of limits to growth. Economists who are so central in setting the political agenda and in informing the public about what is possible and what is impossible have advanced the view that limitless growth is not only possible but a goal toward which society should aspire. Just how far we have to go to educate economists about basic natural laws can be seen by some comments made at a meeting sponsored by the World Bank on the topic of limits to growth. "The notion that there are limits that can't be taken care of by capital has to be rejected," claimed one leading economist. Another attempted to change the default position by challenging, "I think the burden of proof is on your side to show that there are limits and where the limits are." One final plaintive comment left little doubt about what the economists' real concerns were: "We need definitions in economic, not biological terms."

Ultimately, not even the most blatantly arrogant will be able to ignore basic laws of nature. The fact that we have had the creativity to develop the technological capabilities to permit us to fly does not mean that we have overcome gravity, a lesson made abundantly clear with each tragic plane crash. Similarly, the fact that the human population has been growing exponentially for so long does not mean that we have overcome the limitations imposed by the earth's carrying capacity. It is far better to learn that lesson deductively than to wait for the inevitable population crash to teach it to us.

Many economists have so terribly misunderstood the importance of controlling human population growth because of the specialized worldview that they have brought to the subject. Educational efforts, however, are beginning to pay off. Indeed, in 1988 the International Society for Ecological Economics was founded and began publishing *Ecological Economics,* a journal devoted to a more holistic view of economics than had previously been the case.

With some limited progress being made in educating people about the importance of population size, and with serious environmental havoc all too common, you might think that we should have made significant inroads toward achieving a broad understanding about some of the global consequences of our use of technology. In one sense we have. Many more people today than ever before consider themselves environmentalists. But because of a lack of understanding or a misunderstanding of many basic scientific principles, such environmentalism does not run very deep, and people are incredibly susceptible to disinformation campaigns cast in the guise of science by pseudoenvironmental organizations. There is ample evidence that individuals, corporations, and industries with profits at stake are willing to spend huge sums of money to deceive the public into a false sense of environmental security.

Consider one blatant example—the creation of a new "environmental" group. Environmental organizations usually begin as shoestring operations with a good deal of after-hours volunteer labor working secondhand equipment in cramped office spaces or in the living rooms of people's houses. A group calling itself the Information Council for the Environment (ICE) sprang onto the scene in a very different fashion. From the beginning it boasted a toll-free telephone number, a Washington, D. C., public relations firm, and a budget in excess of half a million dollars. None of this was good news for the environment. The Information Council for the Environment, you see, was funded by a collection of electric energy and coal companies, from which its officers are drawn. Its mission, as explained by Gale Klappa, who is both president of the council and vice president of the Southern Company (the parent company of five major electric utilities in the Southeast), is quite simple. The council was formed to educate the American people about the greenhouse effect. Or, more to the point, the council was formed to counter the growing body of information indicating that unless action is taken very soon, serious environmental consequences will result. Klappa and the council believe that the public has been exposed to only one side of a complex issue.

To counter what it perceives as a bias, the council has undertaken a sophisticated media blitz. Its full-page newspaper ads and numerous radio spots claim that catastrophic global warming is not a reality and that the scientific community itself is deeply divided on the issue. The ads present selected anecdotal data showing average temperatures actually dropping in some locations and then call for more research before any action is taken.

The electric and coal companies, under the guise of the Information Council for the Environment, are working very hard to generate a controversy where virtually none exists. The scientific community has reached a greater degree of agreement of the issue of global warming than on virtually any other environmental concern. The World Meteorological Organization and the United Nations Environment Programme established the Intergovernmental Panel on Climate Change (IPCC) to assess the available scientific knowledge on climate change.

The IPCC working group consisted of 170 scientists from twenty-five countries. Drafts of their report were reviewed by an additional 200 experts. The final report thus calls itself "an authoritative statement of the views of the international scientific community at this time."

It is impossible to misinterpret the IPCC report. The world's scientists have gone on record saying that current practices "will result in a likely increase in global mean temperature of about 1 degree C above the present value by 2025 and 3 degrees by the end of the next century." Similarly, the scientists claim that sea level will continue to rise because of greenhouse warming. "The predicted rise is about 20 cm in global mean sea level by 2030, and 65 cm by the end of the next century." The IPCC report notes that these effects will not be uniform worldwide: although global average temperature will increase, it will not be surprising to find local regions showing a temperature decline. The scientists also clearly point out that our knowledge is far from perfect. Further study *is* desperately needed. Nonetheless, they conclude that immediate action must be taken. A recent report from the National Academy of Sciences concludes "that even given the considerable uncertainty in our knowledge of the relevant phenomena, greenhouse warming poses a potential threat sufficient to merit prompt responses."

Yet the Information Council for the Environment claims that the data are incomplete and that there is no evidence for "catastrophic global warming." But catastrophe is a relative term. Perhaps a 65-centimeter (that's more than 2 feet!) increase in sea level is not a catastrophe to the council's executives who have offices on the upper floors of high-rise office buildings, but you can be assured that those of us on the ground floor have much to worry about. Even a 1-foot rise in sea level will move most of the shoreline of the Atlantic and Gulf coasts inland about 100, and in some areas as much as 1,000, feet. Flooding problems will be greatly escalated, freshwater wells will be infused with salt water, and many coastal marshes essential to bird and fish populations will be drowned. Problems in the developing world will be even more dramatic. The densely populated, fertile regions of Bangladesh will be under water. With just a 1-foot rise in sea level, that country stands to lose 10 percent of its land mass and be even more susceptible to the deadly effects of typhoons.

Alice LeBlanc, staff economist at the Environmental Defense Fund, believes, as she stated in a phone conversation, that the Information Council for the Environment is "a clear-cut case of a vested interest sticking its head in the sand and refusing to accept a growing scientific and political consensus." More than that, it is a cynical disinformation campaign designed to confuse an unsuspecting public in the name of education. One of the things that makes such a campaign so morally repugnant is that the consequences of our ignoring loud and clear scientific warnings about global warming fall unevenly on the people of our planet. Consider how much greater the environmental outcry would be if we thought that an increase in worldwide temperature of a few degrees might bring an end to our entire society. Under that sort of threat, it seems unlikely that the possible economic downside of reducing emissions leading to global warming would be paid much heed.

For a number of nations, such a threat is alarmingly real rather than merely hypothetical. Representatives of the governments of fourteen countries have been trying to gain the attention of the developed world and convince it that their very existence is being threatened. For each of their countries, global warming has a particularly frightening ring because of their fragile geographic location. The fourteen nations represented are all island states, small countries scattered throughout the Caribbean, the South Pacific, the Mediterranean and the Indian Ocean. All of these nations have at least one fear in common: rising sea levels might literally pull their societies under water.

A conference to discuss exactly this issue was convened in Male, the capital of the Maldives, by Maumoon Abdul Gayoom, the Maldivian president. South of the Indian subcontinent in the Indian Ocean, the Maldives are typical of the other thirteen island state participants. Comprising more than two thousand small coral formations, none more than five square miles in area, the Maldives have an average elevation of five to six feet above sea level. The stakes for these countries are exceedingly high given the estimates of sea-level rise offered by the IPCC report.

In an impassioned speech, Gayoom said: "Neither the Maldives nor any small island nation wants to drown. Neither do we want our lands eroded or our economies destroyed. Nor do we want to become environmental refugees. We want to stand up and fight. All we ask is that the more affluent nations, and the international community in general, help us in this fight." Another speaker clearly outlined the plight of these nations: "Small island states are in the invidious position of being among the major victims of man-made changes to which they have contributed virtually nothing. . . . Their future survival depends on a greater sense of responsibility by the rest of the world."

As members of an environmentally profligate society, we have an obligation to educate ourselves to ensure that we are not duped by the sort of self-serving disinformation campaigns advanced by groups such as the Information Council for the Environment. We also have an obligation to demand that our government lead an immediate worldwide fight against global warming. As the Maldivian participants pointed out, the issue is not one of mere academic interest. The very survival of whole societies, such as that of the people of Kiribata's Kiritimati Island, known in the Western world as Christmas Island and originally colonized almost seven hundred years ago, is in jeopardy.

It is somewhat odd that there has been so much controversy. The greenhouse effect, or the heating of the earth's atmosphere by radiation trapped by gases such as carbon dioxide, nitrous oxides, chlorofluoro-carbons, and methane, has been understood since late last century. The process is entirely natural and has been responsible for maintaining a relatively constant temperature on earth for millions of years. For more than three decades scientists have warned that problems will arise as the volumes of these insulating gases swell. And swell they have. Ever since the start of the industrial revolution, the amount of carbon dioxide in the atmosphere has risen regularly. Atmospheric carbon dioxide is up about 23 percent since 1750, with an almost unbelievable increase of approximately 8 percent between 1958 and 1980. Worldwide temperatures have risen as well during this period, with four of the hottest years on record occurring within the past decade.

The predicted rise in sea level is due to the melting of the Antarctic ice sheets and the expansion of oceanic water, which increases in volume as it warms. At the same time that the oceans increase in size, higher temperatures causing faster rates of evaporation will lead to a decrease in size of freshwater lakes. Thus, at a time when the warmer temperatures will promote frequent, severe droughts, less fresh water will be available for irrigation. In addition, because increased carbon dioxide often stimulates plant growth but does so at different rates for different species, competitive relationships among species might well change. Agricultural harvests of corn, for example, are expected to lose out to wheat and soybeans. Again, those in the developing countries have the most to lose. Already, Asia's rice production and Mexico's potato production have begun to drop after three decades of continuous growth, and many fear that a warmer, drier climate will exacerbate those trends.

Although it is impossible to predict exactly what might happen in natural communities, plant ecologists are convinced that many species will become extinct and others will increase in prominence. Ecologists are most concerned about the fate of the world's tropical rain forests, whose species-rich habitats are particularly susceptible to irreparable damage. A modest 1.66°C increase in temperature would indeed create a world different from the one in which we currently live. Put this in perspective by considering that during the last ice age, worldwide temperature averaged only about 2.2°C cooler than today.

As with most environmental problems, there are steps that might help, if only we would take them. Since a large portion of the increased atmospheric carbon dioxide comes from the burning of fossil fuels, the most obvious solution is to reduce our reliance on those fuels. A concerted effort to improve our energy conservation habits would certainly help, as would an immediate financial commitment to the development of solar technologies.

Fossil fuels are not the only cause of increased atmospheric carbon dioxide; massive amounts of worldwide deforestation are also culpable. We now destroy about 27 million acres of forest per year, cataclysmic on a planet whose great forests serve as sinks for carbon dioxide. As the trees in those forests photosynthesize, they remove large volumes of carbon dioxide from the air and replace it with oxygen. Obviously no such replacement occurs when these forests are decimated. Additionally, the burning associated with most tropical deforestation releases enormous amounts of carbon dioxide into the air. And much of what is not burned rots, also with significant carbon dioxide emission. The rate at which we are destroying our forests, therefore, needs to be greatly slowed, if not halted completely.

The tragedy is that, aware of the problem for decades, we may have waited too long to act. Scientists are of a mixed mind concerning the effects of those actions proposed above. Some feel that such actions might greatly reduce the anticipated temperature increase, but others believe that the damage is done and that the best we can hope for is a slight delay in its consequences. Regardless of which faction is closer to the truth, we need to begin serious planning for a life under altered environmental conditions. Such planning and action will not be cheap.

It is worth noting that the Netherlands, long fighting an ongoing battle with the sea for control of low-lying lands, spends a larger percentage of its gross national product in defense of its coastline than does the United States on its military. One of the lessons we need to learn from all of this is that it is often much cheaper and easier to stop a global environmental problem earlier rather than later.

That same lesson is readily apparent in the situation regarding the ozone layer, the thin stratospheric shield defending all life on earth from the mutagenic effects of solar radiation. Year by year disquieting reports indicate that the ozone layer is becoming progressively thinner and that the actual holes discovered a while back are becoming larger. The correlation between increased frequency of skin cancer in Australia and the size of the hole over the Southern Hemisphere is a real cause for alarm. What is truly unsettling is that even if we instantaneously halted the production and use of every chemical known to destroy ozone, the holes in the ozone layer would continue to enlarge for the foreseeable future. The reason is quite simple. The chlorofluorocarbons (CFCs) responsible for the destruction of the ozone layer serve as catalysts that interact repeatedly with ozone molecules, destroying many before returning to earth. (In fact, in the upper atmosphere the CFCs break down, releasing atomic chlorine, which reacts with ozone $[O_3]$. This reaction produces the common form of oxygen $[O_2]$ while releasing the chlorine to react with another ozone molecule. Estimates indicate that a single chlorine molecule is capable of destroying approximately 100,000 ozone molecules.) Had serious action been taken back in 1974 when the link between CFCs and the depletion of the ozone layer was first demonstrated by chemists Sherwood Rowland and Mario Molina, most of our current problem would have been averted. As with so many environmental difficulties, however, greedy corporations, an incredulous scientific community, and nervous politicians conspired to preclude the taking of decisive action.

It was so unbelievable that humans could have such a dramatic effect on a worldwide phenomenon like the stability of the ozone layer that scientists actually dismissed the early data as anomalous readings by errant pieces of equipment. When the unbelievable became sadly believable, industry went on the offensive and acted before politicians had a chance. At the first hint of serious trouble in 1974, Du Pont, the world's largest CFC manufacturer (CFCs were originally invented by a Du Pont chemist in 1928 and marketed under the trade name Freon), took decisive public action. Immediately, in a promise made before Congress and in a massive newspaper advertising campaign, the company vowed to "stop production of these compounds" if research demonstrated that CFCs could not be used without threatening public health.

Such unambiguous action had its intended effect. According to Senators Robert T. Stafford (R-Vt.), Max Baucus (D-Mont.), and David Durenberger (R-Minn.), Congress put aside its concerns about regulating ozone depletion because of Du Pont's promise. In 1988, after the link between CFCs and the Antarctic ozone hole had been fully accepted by atmospheric scientists, and after scientific evidence indicated the beginning of ozone damage over temperate areas as well, these three legislators wrote to Du Pont urging that the company keep its word and stop production of CFCs: "We believe the time has arrived for the Du Pont Corporation to fulfill that pledge." Du Pont's response was as unambiguous as was its original guarantee. Du Pont's then chair Richard Heckert called the request "unwarranted and counterproductive."

Needless to say, the saga does not stop there, and although on the surface things might appear to be looking up for the environment, reality is quite different. The two pieces of news which permit naive optimists to dismiss the problem are the international signing and enacting of the Montreal Protocol on Substances That Deplete the Ozone Layer and Du Pont's claim to be producing an ozone-friendly CFC substitute. Du Pont began a media blitz announcing the availability of its CFC replacements. In one advertisement, the company answers the query of a forlorn penguin floating on a block of ice asking, "What on Earth is the future of refrigerants?" by replying, "Du Pont Suva refrigerants. Alternatives to CFCs that are environmentally enlightened. Practical. And available starting today."

So what's the problem? One problem is that Suva refrigerants, though better than old-style CFCs, are far from "environmentally enlightened." Some still use chlorine, the most potent ozone-depleting chemical, and all have significant heat-trapping properties that contribute to global warming. But the greater problem is that Du Pont has been waging an expensive and deceptive campaign against an alternative to its products, even though that alternative is much more environmentally sound.

Although the names look like alphabet soup, they are worth keeping track of. Du Pont's Suva refrigerants are abbreviated as HFC 134a and HCFC 22; the alternative is HFC 152a. Two basic sets of points are not in dispute. First, the alternative, HFC 152a, is 8 percent more energy efficient and, in terms of global warming, 10 percent less harmful than are Du Pont's products. For these reasons, HFC 152a is favored by the EPA. Second, the two Suva products will be immensely more profitable to Du Pont than will HFC 152a. Du Pont has constructed a $30 million plant and committed itself to a $100 million addition for these products and would hate to lose these investments. Even more important, however, is the fact that HFC 152a is a "commodity chemical," meaning that it is in the public domain and available for manufacturing and marketing by any company. HFC 134a and HCFC 22, on the other hand, are patented, exclusive products of Du Pont.

What is in dispute is the safety of HFC 152a. Du Pont has created a video purporting to demonstrate the grave safety problems associated with that chemical. This video, which has been shown to any governmental regulator and representative of the refrigeration industry willing to watch, shows HFC 152a exploding. In one scene a fifty-five-gallon drum filled with the chemical blows up; in another, balls of flame are shown shooting from a refrigerator. The point Du Pont wants to impress on watchers is that HFC 152a's extreme volatility makes it too unsafe for everyday use. What the videos fail to show, however, are the special effects behind the scenes. The drum of HFC 152a ignited not because of its inherent instability but because 15,000 volts of electricity were run through a spark plug built into the specially designed drum. Independent testing firm Underwriters Laboratories engineer Donald Grob explained, "[The drum was] designed to make sure you ignite it. You don't have a 15,000 volt arc in your kitchen." The refrigerator caught on fire because it, too, was specially modified. HFC 152a was pumped into a closed refrigerator containing a bulb with its filaments exposed, peculiar additions to the usual milk and vegetables. In short, although Suva refrigerants are environmentally sounder than standard CFCs, they are not nearly as good as the less profitable HFC 152a. What has become depressingly clear is that we cannot depend on either CFC replacement, even coupled with the Montreal Protocol, to protect the ozone layer from further destruction. The reason, in large part, returns us to the issue that began this chapter—human population growth continues unabated.

The Chinese government has decided to increase its production of CFCs by ten-fold in the coming years. Before condemning the Chinese for their antienvironmental attitude, however, it is worth looking a bit more closely at some specifics. The production statistics tell a very interesting story. As of the late 1980s the United States was producing approximately 1 kilogram of CFCs per person per year; China was averaging about two hundred times less: 0.005 kilograms per citizen per year. The Montreal Protocol compels developed countries such as the United States to cut their CFC production in half by mid-1998 (with a maximum allowable amount of 2.5 kilograms per individual) but allows developing countries that have signed the agreement to increase their CFC usage while placing a cap at 0.3 kilograms per person. China's increased production schedule will still place it well beneath the cap for developing countries. Even if the United States achieves its 50 percent reduction and China attains its goal of a ten-fold increase, on a per capita basis China will still be producing one-tenth the amount of CFCs as will the United States.

The projected Chinese increase in production stems from three factors. First, the huge Chinese population continues to grow at a rate that has exceeded predictions. Current estimates forecast 1.35 billion individuals by the year 2000, up from approximately 1.16 billion of today. Second, China currently is undergoing dramatic modernization. According to Amory Lovins of the Rocky Mountain Institute, "During a five year period in the early 1980s, the percentage of Beijing households with refrigerators rose from two to three percent to more than 60 percent." And reports suggest that refrigerators are still one of the most widely sought consumer items in China. Finally, the status of Chinese manufacturing makes it extremely unlikely that the country will be able to make widespread use of some of the new, environmentally safer CFC substitutes at any time soon. Some of those replacements, particularly Du Pont's HFC 134a, cannot be used in current refrigeration systems but await the spread of newly designed equipment. Italian and Japanese companies, while upgrading their own refrigerator-making equipment to make use of the new chemicals, have been selling China their obsolete gear.

Despite the environmental degradation that it might cause, China's decision to increase its production of CFCs is a hard one to dispute. How can those of us comfortably living in the developed world demand that developing countries not strive for a standard of living comparable to ours? In fact, although China seems willing to abide by the conditions set forth in the Montreal Protocol, the Chinese have, in large part, refused to sign it because it treats the developing countries as second-class world citizens by imposing harsher standards on them than on the developed nations.

Global environmental problems in a world populated by countries at vastly different stages of economic and technological development are proving exceedingly difficult to solve. Real environmental progress will be achieved only when the developed countries begin to design and export technology appropriate for existing conditions in the developing world, even if such projects are quite costly, rather than imposing different environmental standards. Unfortunately, the Reagan and Bush administrations moved us quite a distance from this goal. George Bush, over the strenuous objections of William K. Reilly, administrator of the EPA, refused to provide financial aid to developing countries to help them phase out the use of ozone-destroying chemicals. One proposal called for the spending of approximately $15 million to help the Third World modernize its patterns of CFC use. Although that price tag might seem large, consider that $15 million annually is only 0.9 percent of the total excise tax that the U.S. government has imposed on the domestic CFC industry. One additional issue is critical to bear in mind. Each CFC molecule released at ground level takes up to fifteen years to reach the stratosphere, where it will begin to destroy ozone molecules. Thus, the ozone destruction that is now occurring is the result of our decisions fifteen years ago. Similarly, the decisions we make today will not have a noticeable effect for another fifteen years. As with global warming, we have set in motion a series of events over which we have little control and whose effects might well be cataclysmic for our children. As with global warming, the lesson to be learned is that early remedial action is often cheaper and much more effective than any attempt to respond after the damage has been done.

Although global warming and destruction of the ozone layer are perhaps the clearest examples of ways in which our actions can have a global effect, there are many other, more subtle situations that warrant our attention. Serious environmental effects of any number of our actions may be displaced either temporally or spatially. Consider a report released by the United States Geological Survey demonstrating that rainwater around the country is contaminated with carcinogenic agricultural pesticides. In characteristic fashion, the pesticide lobby has pretended that this alarming situation poses neither environmental nor health consequences. Dennis Vercler, spokesperson for the Illinois

Farm Bureau, dismissed the study by saying, "What concern is it that rainwater contains a very small amount of a crop chemical in New York that came from the Midwest if there are no consequences, either from a health point of view or from an environmental point of view?"

No consequences indeed! The international community has known for decades that organochlorines such as DDT and PCB are stored in fatty tissues from which their slow release has been known to cause cancer, neurological disorders, disorders of the immune system, and premature births as well as a host of other maladies. The international community has also known for years that, other than individuals involved directly in chemical accidents, the people most contaminated with many of these deadly substances are those benefiting least from their use and living at great remove from their points of application.

Although it might be difficult to believe, most native peoples of the Arctic have been found to be highly contaminated with deadly chemicals that they have neither produced nor applied. The 450 Inuit of Broughton Island, Canada, seventy miles north of the Arctic Circle, have the unfortunate distinction of suffering the highest levels of PCB of any human population yet studied. The industrial pollutant PCB surely did not arise from any mismanagement or carelessness by the Broughton Islanders themselves, whose largest industry involves the sewing of handmade caribou parkas. The wastes coming from the one-room, prefab building are hardly anything anyone would be concerned about.

No, the PCBs and the multitude of equally deadly agricultural chemicals that afflict these people wafted from the south. Many overused (and often indiscriminately applied) deadly chemicals are fantastic travelers. Whether they are sprayed or spilled, a goodly portion evaporates into the atmosphere to be distributed by the winds, sometimes for hundreds or even thousands of miles. The wastes return to earth in rain and, if they do not enter the food chain, evaporate again and continue their travels. Because the earth's upper winds blow predominately in a northerly direction, these poisons head for the Arctic.

Making matters much worse for the Inuit is the ecological process known as biomagnification. Biomagnification is the process by which minute quantities of chemicals are increasingly concentrated in the food chain. When pesticide molecules rain down upon the Arctic, they are often absorbed by one-celled marine animals at the bottom of the food chain. Marine fish feed on these contaminated animals and are in turn fed upon by marine mammals such as seals. The seals and other mammals often fall prey to polar bears. Polar bears and seals, along with most Arctic mammals, have especially large concentrations of fat that insulates them from the harsh weather and which also stores these deadly chemicals particularly effectively. So although these chemicals start out in minute quantities in the atmosphere, their concentration has been increased about three billion times by the time they reach the top of the food chain. Because the traditional diet of the self-sufficient people of the north is exceedingly rich in animal fat, these consumers are continuously exposed to dangerous levels of a wide variety of chemicals. That an entire way of life, not to mention the people who live it, is in danger of extermination because of chemical use elsewhere is an injustice of cosmic proportions.

Other studies have shown that some previously unimagined properties of pesticides are causing equally serious problems of a temporal rather than a spatial nature. Pesticides have been found to bind to other substances in surprising ways. This newly discovered binding property of pesticides means that pesticides, even those that we stopped using years ago, are going to stay with us and cause health problems for quite a while into the future.

Studies of toxic ethylene dibromide (EDB) performed by Joe Pignatello and Charles Frink, soil scientists at the Connecticut Agricultural Experiment Station, nicely illustrate the problem. From its registration in 1948 through its banning in 1983, EDB was widely used throughout the country as a soil fumigant to control worms attacking plant roots. Now, nearly twenty years since EDB's last application to Connecticut cropland, Pignatello and Frink are finding significant amounts bound up in the soils under housing developments. Given that EDB is carcinogenic and reduces fertility, serious health problems might well continue long after the use of this pesticide was halted. As Pignatello said, "The soil is acting as a reservoir not only for EDB, but for herbicides, solvents and PCBs." Slowly, over long periods, these bound toxins are released into groundwater.

Similarly, the work of Aladin Hassan, a scientist with the United Nations' Food and Agriculture Organization's Vienna office, has shown just how tenacious and dangerous pesticides can be once they bind to other substances. Grains destined for storage have routinely been sprayed with pesticides to protect them from insect attack, with pesticide content assayed before use to be sure that the toxins have either broken down or been washed away. Having developed a more sensitive assay, Hassan found that far larger amounts of pesticides than ever imagined actually remain bound to the grain. Preliminary tests suggest that these pesticide residues are capable of impairing liver function.

Some of our unwise ecological activities in one location are destroying whole ecosystems in others, including one that has become a personal favorite. As an ecologist, my work has taken me to a large number of fascinating habitats, from Costa Rican rain forests to Australian heathlands, from Colorado's alpine meadows to Florida's Everglades. Yet I was totally unprepared for what I saw when I first entered the waters above the Great Barrier Reef in Australia. My breath was literally taken away by the sight of the magnificent coral and thousands of colorful fish. Snorkeling over one large staghorn coral stand was like floating over endless New England forests in autumn; a similar excursion through plate coral terrain was like a flight over the strata of a canyon. Throughout, the abundant marine life supported by the vast array of structures defied belief.

Today, coral reefs worldwide are in grave jeopardy. "The reefs, which should be teeming with life, are becoming underwater deserts," warns John Ogden, director of the Florida Institute of Oceanography. The consequences of losing a large number of coral reefs are immense. Economically these reefs often serve as the backbone of entire communities. As tourist attractions, many reefs generate a steady supply of cash, and the teeming marine life of many others provides food and a way of life. Biologically, these reefs, composed of millions of tiny, tentacled animals attached to the calcium carbonate matrix they secrete, serve as a breeding ground for a huge number of marine animals. As the reefs are destroyed, populations of many species, including endangered ones such as giant clams and sea turtles, plummet. The living coral itself protects the shore from the battering action of the ocean, and erosion significantly increases as the coral dies out.

The causes of destruction of these important habitats are varied. Some, like the massive mining of coral for construction material in Sri Lanka or the extensive damage done by the anchors of luxury yachts in the Florida Keys and American Samoa, are obvious. Increasingly in recent years action has been taken to prevent this sort of abuse. Marine scientists are just coming to realize, however, that the biggest threat to the world's reefs is much more insidious and is going to be much harder to control.

The reefs are dying, in large part, because of the presence of excessive nutrients and sediments. The nutrients provide an ideal environment for algae, which outcompete the coral, often strangling it. The sediments work in diverse ways. The small particles deprive the coral of both sunlight and oxygen as well as providing a substrate for disease-producing bacteria. The sediments also abrade against reefs, killing new growth. In extreme situations, sediments have completely buried coral.

As with mining, the local causes of the damage are usually fairly obvious. That the algae destroying much of the coral in John Pennekamp Marine Park, off Key Largo, Florida, is a direct result of the effluents dumped into nearby Dispatch Creek, where for years yacht owners routinely flushed their toilets, cannot be disputed. The creek also drains four golf courses that, together, apply approximately forty tons of fertilizer. The foul water of Dispatch Creek flows out to the Marine Park, where it feeds the algae.

Reefs in the Caribbean, suffering from similar types of local abuse, also suffer additional problems arising from environmental degradation occurring hundreds of miles away in South America. Deforestation in southern Venezuela, near the headwaters of the massive Orinoco River, has dramatically increased the amount of erosion occurring. The composition of the Orinoco has changed markedly, and reefs as far away as the Caribbean are being devastated by the material being carried by the river into the ocean. In fact, the aquatic plumes of nutrients and sediments are not unlike the airborne radioactive plumes spewed forth by the Chernobyl reactors.

That reefs are dying because of large-scale environmental destruction hundreds of miles distant reinforces the basic ecological principle that all living things are interconnected, a situation analogous to North American songbird populations plummeting because of tropical deforestation. Such long-range interactions make taking effective remedial action extremely complicated.

When Marshall McLuhan said that we live in a global village, he was thinking primarily in terms of our ability to communicate instantaneously with one another around the globe. What has become painfully obvious is that because of the vast numbers of people on the planet and because of our growing technologies, physically the earth actually has become a global village. As is true for any village, we need to care for all segments of our community, and we need to recognize that actions taken in one part of the village will be felt at the other end of town. Similarly, we need to realize that when we damage part of our village, it is likely that our children will have to pay to correct our mistakes.

Metaphors are only useful up to a point. The one most often used to describe the interconnectedness of life on earth is the "butterfly effect" arising from a developing scientific field called Complexity. Simply put, the butterfly effect says that the flapping of a butterfly's wings over the Amazon rain forest leads to a thunderstorm over San Francisco, an exaggeration whose implicit lesson is one that we would do well to heed carefully. All of our environmental actions have consequences, some very far removed either temporally or spatially from their original locus. We ignore those consequences at great risk to ourselves, our children, and the biotic community of which we are a part.

Chapter Six

ॐ

Endangered
Species

A basic law of ecology is that living things are tightly dependent on one another, often in ways not easy to imagine. Who, for example, would have predicted that when the last dodo was killed in 1675, that death would lead to the slow extermination of the tambalocoque tree, whose hapless seeds germinate only after passing through the dodo's digestive tract? Now, no natural stands of tambalocoque younger than three hundred years are found anywhere. Or who, when we began clear-cutting our tropical rain forests, would have predicted that the felling of trees would so significantly alter local weather patterns that the tropical rain forest biome itself, along with its vast diversity of life, might not survive?

Interactions of this sort are well worth noting not only in their own right but because of the possible ramifications of a phenomenon that ecologists have just begun to document. Mushrooms worldwide appear to have entered a catastrophic state of decline. Throughout Europe, in countries with terrains as diverse as the Netherlands, Germany, Austria, Poland, and England, wild mushrooms are becoming increasingly difficult to find. And those fungi that *are* being found are significantly smaller than those of years ago. Preliminary data suggest that the same troubling situation is occurring throughout North America as well, with the decline so precipitous that biologists have begun to refer to the circumstance as a mass extinction.

Two obvious explanations for the demise of the mushrooms—habitat destruction and overpicking of edible types by an ever growing and eager human population—have been ruled out. Sophisticated sampling schemes designed by ecologists have been able to control for the fact that there is less land available for wild mushrooms; wild mushrooms have been declining at a rate far exceeding that at which land is being developed. That the decline has affected both edible and inedible mushrooms equally indicates that gourmands in search of tasty treats are not the main cause of the problem.

A look at how those fungi interact with other life forms reveals that a worldwide loss of wild mushrooms might not seem like that big a deal, but the consequences may well be very grave indeed. Ecologists fear that if the mushrooms die off, our forests may not be very far behind. Many of the most endangered mushrooms live in extremely close association with trees; the mushrooms provide the trees with water and minerals, and the trees supply the mushrooms with carbohydrates. In addition, the mushrooms' underground filaments often extend much deeper into the soil than do the roots of trees, thus making available to trees resources that would otherwise be lost. Ecologists have found that trees growing in the absence of mushrooms are significantly more susceptible to environmental stress than are those growing alongside the fungi. The situation is so serious that Eef Arnolds, an ecologist from the Agricultural University of the Netherlands specializing in mushrooms, thinks that "severe frost or drought could lead to a mass dying of trees" (*Science,* 6 Dec. 1991, 1458).

Although the cause of the decline has not been pinpointed with certainty, most experts believe that the mushrooms are responding to abnormal atmospheric levels of nitrogen, sulfur, and ozone. Arnolds claims that in the Netherlands the main culprit appears to be excessive nitrogen applied as fertilizer to agricultural fields. Once again we are seeing the unpredictable effects of our own carelessness.

If the experts are correct about the cause of the decrease in mushroom populations, then the mushrooms might provide us with some critical information and insight. Much like the canaries that miners used to carry deep into mine shafts, so that the death of the bird might warn of a lack of breathable air, these other small, indicator species are warning us about the state of our planet.

Mushrooms are by no means unique in either of these respects. Many species, if we are attuned to the messages they send, can serve as indicators of environmental health. A growing body of work demonstrates how lichens can be used as an indicator of air quality. Years ago ecologists recognized the centrality of the northern spotted owl in the Pacific Northwest's old-growth forests. The status of the owl told scientists a good deal about the well-being of coexisting species. As the spotted owl goes, for instance, so go the northern goshawk, Vaux's swift, the silver-haired bat, the red tree vole, and the northern flying squirrel. Even more obviously, the message implicit in Rachel Carson's *Silent Spring* was that if our abuse of pesticides was so decimating songbird populations, then we needed to think critically about what those same pesticides might be doing to the human population as well.

Additionally, because of tight coevolutionary interactions among many species, ecosystem function and health can be greatly disrupted by the extinction or population decline of even a single species. A classic example demonstrating how members of a single species might influence the structure of an entire community comes from experiments performed by Robert Paine, an ecologist at the University of Washington. Working in the rocky intertidal region of the Pacific coast, Paine found stable intertidal invertebrate communities dominated by fifteen species of animals including starfish, bivalves, limpets, barnacles, and chitons. When Paine removed the starfish, the community collapsed until only eight invertebrate species were common. Although it was not obvious in undisturbed intertidal regions, the starfish were preying heavily on one of the bivalves and keeping its numbers quite low. With the predator removed, the bivalve population increased and, because the bivalve is such a good competitor, was able to outcompete many other species of invertebrates. Loss of one species thus rather quickly led to the disappearance of another six as well.

For economic as well as for aesthetic reasons, protection of threatened and endangered species is critical. Because no economic value can be easily placed on the aesthetic value of wildlife and wilderness, the aesthetic reasons for preservation are far too often ignored. This is a sad state of affairs because, after all, part of our own dignity as humans stems from our ability to see beauty and value in other living things. Whether I ever see a northern spotted owl, an Ozark big-eared bat, or a Puerto Rican parrot, I feel better knowing that some of each are still alive. Similarly, just knowing that tracts of true wilderness still exist in North America, whether or not I ever have the good fortune to hike through some of it, is good for my spirit.

Voices far more articulate than mine have explained how we lose a bit of our humanity when we allow species to become extinct. Harvard biologist and two-time Pulitzer Prize winner E. O. Wilson wrote in the book *Biophilia*, "To the degree that we come to understand other organisms, we will place a greater value on them, and on ourselves." Aldo Leopold, as he has done so often, captured the essence of the matter when he wrote, "The last word in ignorance is the man who says of an animal or plant: What good is it?"

At times our greed conflicts with our desire to save rare animals and plants, placing us in distressing situations that force us to make decisions akin to Faustian bargains. Consider a strategy designed by conservation biologists to save the highly endangered black rhinoceros from extermination by poachers. In Damaraland, a semiarid portion of Namibia, rhinos were hunted by biologists in helicopters. The rhinos were shot with a strong sedative before biologists sawed off their horns and administered an antidote to the sedative. The hope was that this extreme measure would relieve the pressure from poachers who have decimated the black rhinoceros population in order to acquire its horns for sale on the black market.

The case of the black rhino is particularly sad because, unlike that of so many endangered or extinct animal and plant species, the major threat to its continued existence is not habitat destruction by humans. Rather, this protected animal, like the elephant hunted for its ivory, is hunted solely for its horns. Those horns are put to a wide variety of odd uses. In the Middle East, particularly in Yemen, the horns are especially valued as a material for the fashioning of handles on traditional daggers. In the Far East, the horns are ground into a powder and prized as an aphrodisiac, as a tonic for headaches, liver ailments, and heart problems, and as a skin ointment. Not surprisingly, modern scientific testing has been unable to substantiate the utility of any of these applications. Nonetheless, whole horns may bring well in excess of $30,000 each on the black market.

Also not surprisingly, such prices are wreaking havoc on the rhinos. Although estimates vary somewhat, most experts agree that approximately 65,000 black rhinos have been shot in Africa since 1975, bringing the total number remaining alive down to somewhere between 800 and 3,500. Whatever the exact numbers, David Western, director of the New York Zoological Society's Wildlife Conservation International, has said, "The black rhino will be extinct in the wild in five years unless we can halt its population decline" (quoted in *New Scientist* [spring–summer 1989]).

Stopping the decline has not yet been possible, even under the best of circumstances. Kenya banned big-game hunting in 1977, recognizing that, as a nation, it stood to make significantly more money through photographic than through hunting safaris. The Kenyan government established 22,500 square miles of national parks and game preserves, including an expensive 700 square miles of electrically fenced rhino sanctuaries. Yet from May 1988 through February 1989, six rhinos and six rangers were killed by poachers in Kenya alone. No one should be misled into believing that the poachers are tribal hunters practicing traditional rituals. They are killers armed with highly sophisticated weaponry—AK-47s, large-caliber machine guns, and even rocket-propelled grenades. As the numbers suggest, they are willing to attack and even kill armed rangers guarding the rhinos. Reports suggest that South Africa has been a major player in the black market, trading arms for horns. Members of the conservation community have also indicated that the government of Taiwan has been particularly lax in enforcing endangered species legislation.

This latest initiative from Namibia, then, the dehorning of individuals, is, in some senses, a welcome enterprise. It must be pointed out, however, that it is unlikely to have any widespread application. Because the rhino uses its horn in self-defense, dehorning can only be attempted in those uncommon places where the rhino has few natural enemies. More important, however, we must ask, what exactly is it that we are saving when we preserve a population of defaced black rhinos? Along with their horns, we have removed virtually all traces of dignity from these magnificent animals. And as we are forced to take such drastic step to protect the remaining members of a vanishing species from our greed, we lose a good deal of our own dignity as well.

The economic justification for the protection of endangered species is as powerful an argument as any of the reasons yet advanced. To many, because of its utilitarian overtones, it is much more powerful. From the cancer-fighting properties of the taxol present in the bark of yew trees through the anticoagulant discovered in leech saliva to the high-protein meat of the Amazon river turtle, there is an amazing array of uses to which we can put natural products. Furthermore, what we are discovering almost daily is that many of the herbal remedies used for centuries by indigenous peoples are based on sound science. Consider the loss for humanity if either the large frog, *Phyllomedusa bicolor*, from the Amazonian basin or the neem tree, native to India and Burma, were allowed to become extinct.

The Matses Indians of Peru, as well as a number of other indigenous tribes, have, for generations, distilled a concoction of secretions from *Phyllomedusa bicolor* and called it "hunter magic." First, an area of skin is ritualistically burned. Then, after this concoction is rubbed on the burned area, a whole host of physiological effects result. Vomiting, defecation, urination, drooling, and profuse sweating as well as greatly elevated pulse rate are induced almost immediately. An ensuing day-long listlessness is then followed by a feeling of transformation. Peter Gorman, an amateur anthropologist who experienced hunter magic in Peru, described the negative effects by saying, "I was hoping and praying that I would die" (*Science*, 20 Nov. 1992, 1306). After the day of listlessness, however, "when you wake . . . you feel god-like." Gorman and natives alike feel as if their senses have been dramatically sharpened and their strength enhanced.

Whether or not hunter magic really improves hunting skills has yet to be investigated, but a paper in the prestigious *Proceedings of the National Academy of Sciences* (14 Nov. 1992) reported on a chemical analysis of the frog secretions. It turns out that there are at least several dozen peptides (strings of amino acids) that are biologically active in the substance. One of those peptides, according to biochemist Michael Zasloff, founder of Magainin Pharmaceuticals Inc., "represents what appears to be a new class of pharmacological mediator." This peptide acts on the chemical receptors for the ubiquitous biomolecule adenosine which are widely distributed throughout mammalian brains. Given that research on these receptors is actively being pursued for a huge range of ailments, from depression, stroke, and seizures to cognitive loss in illnesses such as Alzheimer's disease, the possibility is very real that this thirty-three-amino acid peptide derived from the secretions of an Amazonian frog could become a drug of major human significance.

Every bit as impressive are the uses to which the seeds of the neem tree can be put. According to a report released in 1992 by the National Research Council (NRC), the research arm of the National Academy of Sciences, neem seed kernels can be used to produce pesticides, antiviral and antibacterial agents, and a very potent spermicide, all using simple technology designed over centuries by villagers in India and Burma. The NRC reports that the pesticide derived from the seeds has been found to have "remarkable effectiveness" against more than two hundred insect species, including mosquitoes and the desert locust. The antibacterial property that has been found to be effective at fighting tooth decay has been developed into a toothpaste currently being sold in Asia. Noel Vietmeyer, the director of the NRC research project, admits that researchers "were just blown away" by the varied uses of the neem products and their efficacy. This fast-growing species has already been planted in deforested areas of Haiti and sub-Saharan Africa in an attempt to make productive use of its amazing properties.

The secretions from *Phyllomedusa bicolor* and the products available from neem seeds are just two of the many examples that could be cited to illustrate the abundant uses of natural products. As species become extinct, we lose the ability to investigate their biological properties. (There is, admittedly, a certain irony in the tallying of "uses" to which an animal or plant might be put when in the same breath it should be understood that the aesthetic value of those organisms should be enough reason to justify preservation.) It is clear that much investigation remains to be conducted. An astonishingly small percentage of the world's biotic diversity has even been cataloged, much less studied. As Peter Raven, director of the Missouri Botanical Garden, and E. O. Wilson wrote in an essay in *Science* (13 Nov. 1992, 1099–1100), "The roughly 1.4 million species of living organisms known to date are probably fewer than 15 percent of the actual number and by some estimates could be fewer than 2 percent."

According to Raven and Wilson, the species that we are just in the process of discovering are surprising in almost every respect. Consider four of the examples they cite in their *Science* paper:

> 1) Eleven of the 80 known living species of cetaceans (whales and porpoises) have been discovered in this century, the most recent in 1991; at least one more undescribed species has been sighted in the eastern Pacific but not yet collected.
>
> 2) One of the largest shark species, the megamouth, constituting the new family Megachasmidae, was discovered in 1976 and is now known from five specimens.

3) During the past decade, botanists have discovered three new families of flowering plants in Central America and southern Mexico; one, a remarkable relict, is a forest tree that is frequent at middle elevations in Costa Rica.

4) The most recent new animal phylum, the Loricifera, was described from the meiobenthos [organisms living in the interstices of sand grains] in 1983; many additional new species in the group have since come to light.

It is striking that this late in the twentieth century biologists are still discovering new animals as large as the megamouth shark and an entirely new animal phylum. (A phylum is the major taxonomic grouping just below kingdom. That phyla are very inclusive is evidenced by the fact that all amphibians, birds, fish, mammals, and reptiles, as well the tunicates and the lancelets, are part of a single phylum, the Chordata.) As exciting as new discoveries of this sort are, they are more than offset by the rate at which species are being driven to extinction. Because we have identified such a small percentage of the extant species on earth, there is great controversy over the exact rate at which we are losing them. Nonetheless, those in the best position to offer the most accurate estimates warn of cataclysmic change. Raven and Wilson claim, "Tropical deforestation alone is reducing species in these biomes by half a percent per year, as estimated by the conservative models of island biogeography. This figure is likely to be boosted many times when the impact of pollution and exotic species is determined and factored in." Dr. Thomas Lovejoy, previously a top Smithsonian official and now a high-ranking assistant to Secretary of the Interior Bruce Babbitt as well as an expert on tropical biology, has warned that "a potential biological transformation of the planet unequaled perhaps since the disappearance of the dinosaur" is about to take place. Similarly, Russell E. Train, the first administrator of the Environmental Protection Agency and now the president of the World Wildlife Fund, alerts us to the fact that "we are talking about a global loss with consequences that science can scarcely begin to predict. . . . The future of the world could be altered drastically." Finally, and perhaps most graphically, is the metaphor offered by Donald Falk, executive director of the Center for Plant Conservation at the Missouri Botanical Garden. "Biological diversity is being destroyed at a faster rate than it's being described. We have a situation where our library of life is being burned at a phenomenal rate—and we have only a small number of people who know how to read what's left."

Dramatic action is clearly called for. One place to begin is to dispel the greatest myth associated with preservation—that it is simply too expensive to preserve habitats or to save many species.

Whenever a discussion of the economic value of tropical rain forests is undertaken, the potential for future medical discoveries is pitted against the current use to which the land could be put. Since most medical advances are both hypothetical and well off in the future, short-term, destructive uses usually win out. But as Michael Balick and Robert Mendelsohn point out, some medical benefits can be derived immediately. Balick, director of the Institute of Economic Botany at the New York Botanical Garden, and Mendelsohn, professor of forest policy at Yale University, addressed this point head-on in a March 1992 paper published in *Conservation Biology*. Balick has been quoted in the *New York Times* (28 Apr. 1992) as saying, "We wanted to identify what is valuable to the small farmer today, because he decides whether to cut his piece of the forest to feed his family or to use it in another way to derive income. For the first time we are not talking about medicinal benefits that are years in the future. We are talking about benefits that people are realizing today."

Balick and Mendelsohn, with the help of a local herb gatherer, collected samples of all the medicinal plants from two small plots of forest in Belize. The plants, which were sold in local markets to pharmacists and healers, could be used to treat a wide variety of conditions including indigestion, rheumatism, colds, and diarrhea. The sale of these medicinal herbs yielded significantly more money than local landowners could generate in any other fashion, demonstrating that significant profits can be made from tropical herbs today. And moving to a program of sustainable harvesting is even more profitable.

The economic myth needs to be dispelled here at home as well. As our Cold War enemies began to fade away, many people started looking for new scapegoats to blame for a sluggish economy. An enormous amount of attention has been focused on environmentalists, whose concern for the nonhuman constituents of our earth has often been accused of limiting the standard of living possible for humans. The Endangered Species Act, in particular, has been attacked unmercifully in recent years as a relentless drag on the economy.

Although opponents of this showcase piece of environmental legislation continually claim that deference to "unimportant" animals and plants has repeatedly precluded development projects, a study conducted by the World Wildlife Fund (WWF) clearly demonstrates that such a view is far from the truth. Using material acquired under the Freedom of Information Act, the WWF tallied the total number of development projects assessed by the U.S. Fish and Wildlife Service for possible negative impact on endangered species during the five-year period between 1987 and 1991. Of the 73,560 projects evaluated, only 18, or about two one-hundredths of 1 percent, were halted because of the risk posed to endangered species. John Sawhill, president of the Nature Conservancy, very nicely put these numbers into context when he said, "In the same period, 29 airplanes crashed into commercial or residential buildings in the U.S. That means that a developer faced a greater chance during that time of having an airplane crash into something he built than having a project stopped by the Endangered Species Act" (*Wall Street Journal*, 20 Feb. 1992). Nonetheless, data of this sort have not quieted critics. Pete Wilson, governor of California, arguing from emotion rather than reason, stated, "To those of us who are trying to survive in California, [the act] is a blunt weapon that costs us jobs. The endangered species that I'm most worried about is the California worker."

Nor can it be honestly said that we have been spending an undue amount of money to administer this act. According to calculations made by the Environmental Defense Fund, total spending by the federal government to administer the Endangered Species Act during its first seventeen years has been less than the Sandia National Laboratory in Albuquerque spent on nuclear weapons research in 1991 alone.

Finally, even given the very modest amount of money devoted to endangered species, it is impossible to argue that the act has not had a beneficial effect on some of the wildlife that it was designed to protect. As of 1995, six species have been removed from the endangered and threatened lists because of the significant rejuvenation of populations. Because of the low level of funding, however, there have been losses as well. As of 1995, seven species have been removed because they have become extinct, and estimates indicate that with the current rate of funding, it will take approximately fifty years to list all of the nation's species currently considered to be in danger of extinction. Another concern is that species are not necessarily being added to the list in a sequence reflecting their probability of extinction. Botanists worry that because endangered plants are less likely to grab the public's attention than are cute animals, our endangered flora is receiving a good deal less attention than our endangered fauna.

Attacks on the Endangered Species Act and on imperiled wildlife itself reflect a terribly narrow worldview that considers only short-term profit and omits virtually every other environmental, economic, aesthetic, and social value. The strident attack on a pesticide labeling program proposed by the EPA is an excellent case in point. The program stemmed from the EPA's recognition that because pesticides are designed to kill organisms, their indiscriminate use might have negative effects on some endangered species. Encouraged by environmental groups, the EPA developed a pesticide labeling program fashioned to protect those rare animals and plants. The labels, which are a model of simplicity, would list those counties in which an endangered species might be threatened by use of the chemical. Detailed maps of endangered species at the county level would be made available, and a particular pesticide could still be used in a listed county if the county agricultural agent determined that its use did not impact negatively on the endangered species.

What made the program so reasonable was its recognition of the necessary balance between the protection of endangered species and the agricultural need for pesticides. Use of specific pesticides would be prohibited only in the particular locations where endangered species might be threatened. Farm groups and the pesticide industry, however, were up in arms over this plan. Their attack came in two forms. The first was a smokescreen centered on the contention that the plan was impractical. Don Rawlins, director of natural resources for the American Farm Bureau Federation, one of the groups leading the opposition to the labeling program, argued that we simply do not yet have enough information to proceed. In a telephone conversation with me, he claimed that endangered species are not often found where they are thought to be and that, in any event, we have little hard knowledge of the effect that specific pesticides might have on them. He also asserted that the government has not fully assessed the economic impact of pesticide bans in various locations. Given that the Endangered Species Act does not mention economic trade-off—endangered species are to be protected from actions that will jeopardize their survival, period—this latter point is easily dismissed. That our knowledge is incomplete obviously calls for a conservative strategy. The burden of proof that pesticides will not disrupt the populations of endangered species should fall on pesticide users rather than on the endangered species themselves. When we are dealing with vanishingly small populations of animals and plants, we simply cannot afford to act only after discovering that our policies have had tragic consequences.

A quick conversation with Rawlins, however, brought him to the second reason for the attack on the program: Rawlins simply does not believe that many endangered species are worthy of protection. "When a species's time has come," he announced, "perhaps there is nothing we should do." He went on to contend that "man may not be perfect, but he is a superior species and he is the one we need to protect." This is really the heart of the matter. Most antienvironmentalists create a false dichotomy between human beings and the natural world and say that the former are more important than the latter.

Columnist Charles Krauthammer, writing in the 17 June 1991 issue of *Time*, took this attitude one step further. He argued that what we must do is "first, distinguish between environmental luxuries and environmental necessities. . . . The important distinction is between those environmental goods that are fundamental and those that are merely aesthetic. Nature is our ward. It is not our master. It is to be respected and even cultivated. But it is man's world. And when man has to choose between his well-being and that of nature, nature will have to accommodate. Man should accommodate only when his fate and that of nature are inextricably bound up" (82).

Ah, if it were only so simple. Far too frequently it is all but impossible to distinguish between environmental luxuries and environmental necessities and between fundamental and aesthetic goods. The price of miscategorization is clearly enormous. More important, it is one thing to demand that "nature will have to accommodate" and quite another to ensure that such an accommodation actually takes place. The lesson that we all need to learn is that we are a part of nature and, regardless of how technologically advanced our society might become, it is impossible to set ourselves above the natural world.

While technological innovations might appear attractive to some, they are no substitute for wildlife preservation. Julian Simon, professor of business administration at the University of Maryland and a frequent critic of environmentalism, wrote an op-ed piece in 1992 crafted to dismiss environmentalists' concerns over the escalating rate of extinction.

He put forward a number of reasons why we should rest easy in the face of the possible extirpation of natural populations. "First, it is now practicable to put samples of endangered species into 'banks' that can preserve their genetic possibilities for future generations. Second, genetic recombination techniques now enable biologists to create new variations of species."

Ignoring the critical fact that Simon's attitude disregards the role that species play within a community, I certainly do not want to live in a world where instead of seeing beauty in the world around me, I must travel to a laboratory to see DNA sequences. And I hope that we never reach the point where my grandchildren or my great-grandchildren find such a prospect any less horrific than I do now. We simply cannot dismiss the aesthetic value of species diversity in the (false) name of economics and expect to retain the remotest shred of our humanity.

We are inextricably linked to the natural world and that's that. Unless we come to our senses quickly, we are left to wonder, with Herman Melville, "whether Leviathan can long endure so wide a chase, and so remorseless a havoc; whether he must not at last be exterminated from the waters, and the last whale, like the last man, smoke his last pipe, and then himself evaporate in the final puff."

Chapter Seven

ॐ

Better Living
through
Chemistry?

"In the minds of the general public, the chemical industry lacks credibility, cannot be trusted, is foot-dragging in its efforts to improve itself, and is neither honest nor ethical in its dealings with others." Those strong words come not from an environmentalist but rather from the Chemical Manufacturers Association, a trade lobbying group. If, in fact, such thoughts are "in the minds of the general public," there is good reason. Over the years, the chemical industry has resisted virtually every effort made by governmental agencies to protect the general public while far too frequently acting as an irresponsible shepherd of its most dangerous products.

The *Charleston* (W.Va.) *Gazette,* after having published a column of mine levying similar charges, ran a rebuttal from an irate reader (25 Mar. 1992). His letter began: "Dr. Michael Zimmerman should be ashamed of himself for fostering chemophobia, which is so prevalent in society today. . . .I am sure that as a man of science, Zimmerman knows that our very existence depends upon chemicals, both natural and manmade. I am also sure he knows that just because a chemical is manmade, it is not inherently dangerous or toxic."

The sexist language aside, it is worth addressing some of the points raised by these few sentences. The letter writer is absolutely correct—all life is dependent upon chemicals, and not all chemicals made by humans are dangerous. Similarly, there are enormous numbers of natural chemicals that are terribly toxic. All of that misses the point, however. Humans have produced, overused, and dreadfully misman- aged a wide array of deadly compounds, many of which, such as insec- ticides, herbicides, and fungicides, were created for the sole purpose of killing. Now don't misunderstand me. I fully recognize that public health concerns necessitate the killing of organisms on numerous occa- sions. The problem is that chemicals designed for killing nonhuman organisms, not surprisingly, often have negative effects on humans as well, especially when such chemicals are used by poorly trained profes- sionals or by a public largely ignorant of possible side effects.

The deadliness of some natural chemicals is wholly irrelevant. Be- cause we cannot avoid encountering a spectrum of natural risks is no reason to increase our exposure to risks we can control. One might even argue that if the health risks posed by chemicals are cumulative and synergistic (and many believe they are), we should be especially careful about increasing an already heavy burden.

What is most interesting is the charge that a newspaper column fosters chemophobia by discussing some of the problems associated with practices of the chemical industry. If a phobia is an unfounded fear, then chemophobia must be an unfounded fear of chemicals. Many chemicals are indeed dangerous, and thus a fear of them is not a pho- bia. As they say, you're not really paranoid if someone is really out to get you.

Just how much do we have to fear from our heavy reliance on chemicals? The chemical industry has been scrambling to rehabilitate its public image while attempting to convince a skeptical and largely ignorant public that there is no reason to be afraid. Elizabeth Whelan, executive director of the American Council on Science and Health, has claimed that "what this country needs right now is a national psychia- trist" to deal with its nosophobia, or morbid dread of illness. Despite the fancy-sounding name of Whelan's organization, it is nothing more than an apologist for the chemical industry. At a meeting of the Na- tional Pest Control Association a while back, she advocated the estab- lishment of a cabinet-level position for national pest control: "I'm talk- ing about controlling human pests, the pests who continue to terrorize us about the quality of our environment. Human pests are damaging the quality of life in America. These people are not promoting health but their own political agenda."

The attempt at rehabilitation of the industry's perception by the public is coming not only from the chemical industry itself; even segments of the broader scientific community are getting involved. Bruce Ames, a prolific and oft-quoted toxicologist at the University of California at Berkeley, has repeatedly attempted to convince anyone willing to listen that we really have very little to fear from mass-produced chemicals. He has gone on record suggesting that the aflatoxin naturally occurring in peanut butter is significantly more dangerous than most pesticide residues found on foods. In a lead article in the prestigious magazine *Science*, Ames claimed that naturally occurring carcinogens in our food supply are more toxic than industrial carcinogens and that the "high costs of regulation" are unwarranted.

One of the most common refrains from Whelan, Ames, and a host of others is that all of our conclusions are misleading because they stem from tests on animals. "A mouse is not a little man," Whelan loves to announce, implying that all of the mouse data are meaningless when it comes to drawing inferences about human beings. Could it be that the plethora of cancer deaths observed in laboratory animals is indeed meaningless? The Environmental Research Foundation, in a pithy little essay published in its weekly newsletter, *Rachel's Hazardous Waste News* (19 Feb. 1992), dealt with this issue by focusing on a very special laboratory animal and asking the provocative question, "What do chemists die of?"

That excellent piece summarizes twenty-three years of international studies examining the causes of death of a myriad of chemists. The results should give even the most vocal supporters of the chemical industry something to think about. A 1969 study published in the *Journal of the National Cancer Institute* examined the causes of death of 3,637 members of the American Chemical Society and found that the chemists suffered disproportionately high rates of cancer of the pancreas and cancer of the lymph system. The former was unsettlingly common in young males of the profession; females experienced elevated levels of breast cancer. Reports from the 1970s demonstrate that not only American chemists are in peril. Swedish chemists sustained high rates of lymphomas, blood-related cancers, and brain tumors, and their British colleagues were found to have surprisingly high levels of a wide variety of cancers, with lymphomas predominating.

Nor does it stop there. A study published in the 1991 edition of the *Archives of Environmental Health* demonstrated that Exxon scientists, engineers, and research technicians had significantly elevated rates of leukemia and lymphatic cancers relative to Exxon managerial employees much less likely to experience chemical exposure in the workplace. Copious studies published from 1978 to 1986 in journals such as the *British Journal of Cancer, Environmental Research*, the *American Journal of Public Health*, the *Journal of Occupational Medicine*, the *Scandinavian Journal of Work, Environment, and Health*, and the *American Industrial Hygiene Association Journal* all reported the same types of patterns.

After reviewing this wealth of distressing material, the Environmental Research Foundation concluded, "There seems to be little doubt that working with chemicals creates a link of cancer, even among those people who are well-educated, who presumably have a healthy respect for the hazards of their workplace, and whose employers are wealthy companies that can afford to take every precaution against excessive exposure." When chemists are exposed to the same chemicals as are lab animals, they develop the same sorts of maladies. Although "a mouse is not a little man," we surely are much better off experimenting on rodents than on humans. That's food for thought the next time someone tries to convince you that industrial chemicals are no more dangerous than peanut butter.

For years, industry's attitude toward its employees with respect to chemical hazards has been backward. Instead of ensuring that all industrial and agricultural workers remain unexposed to dangerous levels of toxic chemicals, common practice has been to decide which workers are most vulnerable to the effects of those chemicals and to exclude them from the workplace altogether. Because the common stance of industrial leaders has been to claim that fertile women are most at risk, or more to the point, that their unborn children are, these women have been systematically excluded from various jobs.

A 1980 survey of chemical companies conducted by the trade journal *Chemical and Engineering News* found consensus in the view that no fertile woman should be permitted exposure to any chemical posing threats to the health of a fetus. Instead of creating a workplace in which employees are free from such exposure, the companies elected either to transfer women to lower-paying jobs, to fire them outright, or to require them to be surgically sterilized. The classic situation is the one in which women at the Willow Island, West Virginia, American Cyanamid plant found themselves in 1978. The company decided that the only women between the ages of sixteen and fifty who would be permitted to hold anything other than janitorial jobs in the plant were those who could demonstrate that they had been surgically sterilized. In an effort to be "fair" to a significant portion of its workforce, American Cyanamid agreed to cover the cost of sterilization for any female employee requesting that procedure.

After numerous court proceedings, the U.S. Supreme Court, in a case involving similar discriminatory practices by Johnson Controls, finally ruled in 1991 that such abhorrent labor practices are unacceptable. There are, the court recognized, other options. Although it might not be cheap, companies could redesign the workplace to ensure less exposure to toxic chemicals. Or they could provide respirators and require that all workers wear them. But it is always easier to victimize half the population than to change manufacturing practices. An American Cyanamid spokesperson, fearing that increased safety requirements would limit productivity, may have unwittingly predicted the future when he was quoted by *Chemical and Engineering News* as saying, "The ideal is that the workplace has to be safe for everyone. . . . In the real world that's totally unachievable without emasculating the chemical industry."

A frightening collection of recent studies has begun to suggest that emasculation is indeed taking place. Medical researchers have conclusively documented two striking patterns. First, although for years many did not want to believe it, we now know that, just like females, males exposed to toxic chemicals are capable of passing birth defects along to their children. Second, we now know that the amount of sperm produced by the average male has declined precipitously over the past fifty years. Epidemiological studies suggest that the most likely cause of this drop in male fertility is environmental, rather than genetic, factors.

Again, *Rachel's Hazardous Waste News* (20 Jan. 1993) has done a real service by publishing a cogent summary of a number of technical studies making it clear that environmental toxins can have very serious effects on sperm. Toxins have been shown to damage sperm directly and have been found capable of entering sperm and being carried into the egg at fertilization, where they cause serious developmental problems to the fetus. The results are terribly disconcerting. A huge Finnish study of 99,186 pregnancies demonstrated a significantly increased likelihood of spontaneous abortion if the father was exposed to toxic chemicals associated with the production of rubber products and solvents used in oil refineries. A separate Finnish study of 6,000 men showed that paternal exposure to organic solvents nearly tripled the rate of spontaneous abortions. The relatively common chemical toluene was found to be a particular culprit.

In addition to spontaneous abortion rates, birth defects are also amplified when fathers have been exposed to toxins. A British Columbian study published in 1990 in the *American Journal of Epidemiology* (vol. 131, 312–21) found dramatically increased rates of heart defects in children whose fathers had been exposed to high levels of carbon monoxide and polycyclic aromatic hydrocarbons. Paternal exposure to chemicals has now also been shown to lead to childhood cancer. Links have been conclusively demonstrated between Wilm's tumor (a childhood cancer of the kidney) and paternal exposure to lead, between childhood brain cancer and a father's work in the petroleum and chemical industries, and between leukemia and paternal contact with paint fumes.

Not only do environmental agents damage sperm, they decrease the amount of healthy sperm produced by the average male. A massive study published in 1992 in the *British Medical Journal* by researchers at the University of Copenhagen found that average sperm counts have decreased 42 percent, from 113 million to 66 million sperm cells per milliliter, within the past fifty years. The study was able to factor out a host of extraneous factors, like the once held view that men's tight-fitting briefs elevate body temperature to the point that sperm production is jeopardized, and concluded that the decrease is due to hazardous chemicals in the environment.

That general conclusion should not be surprising, since previous researchers have found that workers in the chemical and pesticide industries have experienced dramatic drops in sperm production, sometimes to the point of sterility, after chronic exposure to toxic compounds. Sperm counts returned to normal after exposure had been halted. What is novel and truly frightening about the results reported in the *British Medical Journal* is that the decline in sperm production was not limited to men with particularly high occupational risk. Rather, the declines were found in the general population, demonstrating that we have reached the point that ambient levels of toxic chemicals are high enough to affect our health.

Our understanding of the role that toxic compounds play in reproduction has reached the point where we can no longer ignore the problem. When industry officials complain of the high economic cost of cleaning up both the workplace and its by-products, the public needs to be reminded of the high economic cost of caring for children born with severe birth defects and of the increasing costs associated with loss of fertility. Even more important, all of us need to be made cognizant of the human suffering caused by failure to clean up the workplace.

On paper, at least, the chemical industry itself seems to have recognized exactly this point and has recently initiated a massive campaign designed to educate the public. Unfortunately, rather than promoting meaningful education, the campaign attempts to clean up the badly tarnished image of an industry reeling from a host of deadly accidents and lack of public trust. With great fanfare, the industry began what it termed its "Responsible Care" program. *Chemical Week,* the major industry journal, devoted the entirety of its July 1991 issue to this idea. In bold and colorful print, the cover of that special issue of *Chemical Week* proclaimed, "Responsible Care may prove to be one of the biggest and best things that ever happened to the chemical industry. It represents a total, public commitment to continuous improvement of management and performance in the arenas of health, safety, and the environment. . . .With Responsible Care, the chemical industry has embarked on a journey of profound cultural change, opening its doors to a skeptical public and saying, 'Don't trust us, track us.'" The first of ten "Guiding Principles of Responsible Care" goes even further and states that it is the industry's duty "to recognize and respond to community concerns about chemicals and [the industry's] operations."

A study by the U.S. Public Interest Research Group (U.S.PIRG) completed approximately one year after the big splash in *Chemical News* demonstrates just how hollow those words actually are. The study, entitled "Trust Us. Don't Track Us," reports on efforts to gain basic information from 192 chemical facilities in twenty-eight states. All facilities involved in the study are members of the Chemical Manufacturers Association, and thus all have subscribed to the Responsible Care program. Nonetheless, at only 19 of the 192 facilities (9.9 percent) were spokespeople willing to answer all of the basic questions posed by researchers. Despite the industry's supposed commitment to openness, researchers, even after multiple attempts, were unable to reach a single responsible official at 81 (42.2 percent) of the facilities. At an additional 31 facilities (16.2 percent), the contact person refused to answer any questions at all. In other words, at more than 58 percent of the facilities sampled, absolutely no information was forthcoming.

It's not as if answering any of the questions would necessitate divulging trade secrets, not when what was being asked was whether or not the facility has made public its accident risk reduction plans and the routes through the community the facility uses to ship its toxic and hazardous chemicals. No, the information requested is data the public has a right to know and falls well within the industry's much touted Responsible Care guidelines. The U.S.PIRG study clearly demonstrates that the majority of the chemical industry views the Responsible Care program as nothing more than an advertising gimmick. How else is it possible to reconcile the U.S.PIRG study with full-page ads in mass-market magazines claiming, "You're driving by that chemical plant, just like you do every day, when one of your kids asks you what they make in there and you answer that you're not really sure and it occurs to you that you probably should be." The tag line of those ads reads: "The Chemical Manufacturers Association. We want you to know."

"We want you to know" is terribly hypocritical given the wealth of evidence demonstrating that the industry has not backed away from its longstanding objections to informing consumers of basic chemical safety information. Former Congressman Gerry Sikorski (D-Minn.), perhaps the most outspoken member of Congress in favor of fuller disclosure of chemical risks, stated that the industry has repeatedly argued against such disclosure by claiming "it would confuse—even harm—communities by giving them meaningless and unnecessary statistics." Such a paternalistic attitude is disgraceful in any industry but especially so in one that has as abominable a safety record, as does the chemical industry. That that industry then turns around and spends millions of dollars annually on a public relations campaign extolling its openness makes the situation that much more grievously absurd.

Despite the recalcitrance and downright arrogance of the chemical industry, a significant amount of progress has been made with respect to the right of citizens to know what is going on in their communities. One of the most dramatic steps forward in this regard was the Emergency Planning and Community Right-to-Know Act of 1986 (EPCRA). A significant part of this federal act requires industries to report their toxic emissions to both the federal Environmental Protection Agency and to local state agencies. The EPA is to compile this information annually into the Toxic Release Inventory (TRI), making it widely available to journalists, environmental groups, and other interested parties. It was the horror of the Union Carbide accident in Bhopal, India, in 1984, in which thousands of individuals died because of poor planning and poorer communication, which spurred our federal government to action. Even with the Bhopal's carnage as a backdrop, however, the U.S. House of Representatives was remarkably cautious about bucking industry lobbyists. On 10 December 1985 the House passed the first version of EPCRA by the slimmest of margins, 212 to 211.

Although industry lobbyists worked overtime in an attempt to prevent EPCRA from being passed in the first place, those same lobbyists are now quite pleased with the outcome. Former Congressman Sikorski, a driving force behind the original bill, put it best: "They said it was radical. They said it would cost money. It was unworkable, unfair, un-American. Now EPA touts it, Wall Street embraces it, and big companies report they are saving millions of dollars as they cut chemical use."

As wonderful as the Toxic Release Inventory portion of the law is, and given the productive use to which environmental groups have put the data, there can be little doubt that, although it is indeed wonderful, the Right-to-Know Act is far from perfect. A study by Congress's nonpartisan General Accounting Office (GAO; June 1991) points out some very troubling problems. On the positive side the GAO notes that "federal and state governments have used the data to enact laws designed to control and reduce toxic emissions. Also the public availability of the data has prompted some companies to set emissions reductions goals." According to the GAO, the negative side of the ledger, however, is also quite large. One big problem is with compliance. Although reports were submitted from approximately 19,000 facilities, the GAO estimates that at least 10,000 additional plants were obligated by law to report emissions. The GAO concluded that the EPA was "using inefficient strategies to identify nonreporters" and was taking no action against those companies filing late.

The second big problem is with the quality of the data that have been gathered. Shockingly, the GAO report found that EPA officials visited only 27 of the 19,000 sites reporting emissions. Problems, including inaccurate estimates and failure to submit reports for all chemicals, were found at almost one-third (8) of those 27 sites. If that statistic were to hold, more than 6,300 of the reporting sites would be submitting misleading information. In a classic understatement, the GAO report concludes that "if the Toxic Release Inventory is to be used extensively, government regulators and others must be assured that its data are sound," and it goes on to recommend that significantly more effort be spent assessing the quality of the data being reported.

The third big problem discussed by the GAO report centers on what has to be reported and by whom. Currently EPCRA does not require reporting by any federal facility. Although in 1988 the EPA asked federal facilities to voluntarily report their toxic emissions, very few have done so, and EPA has not made a big issue of this. As a result of lobbying pressure during the bill's adoption, EPCRA also excluded "mineral mining and processing, oil and gas extraction, and agricultural operations" from the reporting requirement. These industries, together with federal facilities, produce a significant percentage of the nation's toxic emissions. Equally important is that the EPA requires reporting on only approximately three hundred chemicals. Although EPCRA legislation permits the EPA to expand that list, it has not seen fit to do so.

When sites not in compliance with TRI legislation are added to excluded facilities and unscrutinized chemicals, a huge percentage of the nation's toxic emissions are going unreported. The GAO estimates that "as much as 95 percent of total emissions" may not be reflected in inventory reports. Clearly, the legislative intent of EPCRA has not been met by current practices. And that's the opinion of the General Accounting Office, not of environmental groups that are so often wrongly accused of exaggerating the situation or of promoting chemophobia in the hope of generating additional revenue from supporters.

A number of other problems with EPCRA remain unaddressed by the GAO and have serious consequences for many communities. First, as written, EPCRA has a huge loophole built into it. The reporting regulations for toxic chemicals are different for those chemicals moved off the site at which they are created. Consider the result of this loophole. When a toxic chemical generated by a factory is incinerated on-site, or when that chemical is processed for recycling or reuse on-site, the EPA is supposed to be informed of the amount of the toxic waste originally created. When that same toxic chemical is shipped off-site, either for burning or recycling, no one needs to be informed of the amount of original waste. Because no records are kept, it is impossible to estimate with any degree of certainty how much toxic waste is being driven off-site through this loophole. The amount is clearly not insignificant, however. The EPA estimates that at least 3.6 billion tons of toxic waste are burned annually in industrial boilers and furnaces around the country.

The problem is one of bureaucratic doublespeak. "Toxic wastes," once they are taken off-site, become "industrial products." Calling these noxious chemicals products is especially absurd when in most cases the original manufacturer pays someone to haul these "products" away. But in federal jargon, as long as the substance is going off-site, it makes no difference if someone is buying it or if the manufacturer is paying to have someone else dispose of it.

If the intent of the original law was to enable various governmental agencies to monitor the amount of hazardous chemicals within their jurisdictions, then the off-site loophole undermines that intent. Numerous companies have taken advantage of the loophole to dramatically reduce the amount of toxic emissions they are reporting. The same volume of toxic chemicals is still present—it is still being processed and/or released into the environment—but it now goes legally unreported because the processing and release are being done at separate locations. Using this reclassification ploy, a Monsanto operation in Ohio was able to report enormous reductions in toxic releases. Upon close examination, the Environmental Defense Fund determined that 96 percent of the reduction could be accounted for by a change in terminology rather than by improved environmental stewardship.

The loophole poses several other causes for concern. The current structure of the law encourages large-scale, unreported transport of hazardous wastes. Such shipments pose very real environmental hazards, and because they go unreported, the public has little, if any, opportunity to participate in a debate about their transport through communities. The chemical industry's complaints about EPCRA are indicative of their general feelings about disclosure. No law is unambiguous. Yet every time industry encounters an ambiguous passage, it fights to limit the amount of information that must be released to the public. In spite of all of the pious claims made in the Chemical Manufacturers Association's Responsible Care literature, industry argues for nondisclosure of basic, critical information whenever a gray area is found.

This stance is further reflected in the chemical industry's reluctance to follow the law requiring that it inform local emergency planning committees (LEPCs) of the types and locations of those chemicals.

The law stipulates that, at a state's discretion, the locations should be described either verbally or on a map of the facility. For no rational reason, the industry has been vigorously fighting governmental requests that maps be provided. In Ohio, for instance, the Ohio Chamber of Commerce and B. F. Goodrich have repeatedly challenged the State Emergency Response Commission's regulation that site maps showing the locations of all hazardous chemicals be provided to LEPCs unless an alternative arrangement can be worked out with the local fire chief. Such maps are absolutely essential if any emergency preplanning is to be undertaken. The risks to workers, firefighters, and the general public are magnified enormously when emergency personnel arrive at the scene of a disaster without reliable information. No such information was available on 3 May 1991 when a fertilizer plant exploded in Sterlington, Louisiana. Firefighters gathered outside the facility and, with the aid of workers, desperately tried to sketch the highlights of the plant so they could assess where the next explosion would most likely occur. Eight people were killed and more than one hundred others were injured that evening.

The Ohio regulations, like those being proposed elsewhere, are supported by a broad coalition of trade unions, firefighters, and environmental groups. As Kevin Watts, a district vice president of the Fire Fighters Union, said to me, "a map is as important as knowing what's inside a facility." Nonetheless, industry lobbyists refuse to budge. Sandy Buchanan, a member of the Ohio State Emergency Response Commission and the Toxics Action Director of Citizen Action, a statewide environmental group, believes that the map requirement is being fought so strenuously because the industry fears that capitulation to this regulation will invite more restrictions. Firefighter Watts had perhaps the best perspective when he said that "being a lobbyist in the state capital is different than riding on the back of a fire truck."

The location of hazardous chemicals is not the only piece of information which industry lobbyists have been fighting to keep from the public and from emergency personnel. They have been battling to keep the list of dangerous chemicals for which reports must be filed as small as possible while arguing that the volume of a chemical that would trigger reporting requirements should remain as large as possible.

The controversy stems from the original wording in EPCRA. Congress, when enacting the legislation, recognized that a period of transition was necessary if the huge amount of data was to be organized efficiently. Accordingly, toxic chemicals were divided into two classes; a small minority were termed Extremely Hazardous Substances, while those remaining were simply labeled toxic. While the system was being established, reports would be filed on the latter group only if more than 10,000 pounds of a chemical were present in a single location. After a transition period, information on all hazardous chemicals, regardless of quantity, would have to be shared with those offices responsible for disaster planning. Industry representatives, attempting to limit disclosure responsibility, have been fighting to ensure that the transition period never ends and that the 10,000-pound threshold figure becomes a permanent fixture of the law. Such an event would be a grave threat to public health.

If the exclusion of any chemical from the Extremely Hazardous Substance list gives the impression that the danger posed by that chemical is minimal and thus that a 10,000-pound threshold is reasonable, ask yourself the following questions. Is it reasonable to allow firefighters to enter a warehouse unaware that 9,999 pounds of highly explosive nitroglycerin are present? Is it reasonable to expect emergency personnel to figure out, only after examining dead and dying victims, that huge amounts of methyl chloroform were stored in a building? Neither chemical is on the Extremely Hazardous Substance list. There are, in fact, thousands of toxic chemicals, carcinogens like vinyl chloride and trichloroethylene or explosives like propane, which failed to make the list. Furthermore, imagine what might happen if 9,000 pounds of each of three highly explosive chemicals were stored adjacent to one another. According to the wishes of the industry's lobbyists, firefighters would likely find out about that unfortunate situation only after the resulting chain of explosions.

What possible rationale, then, could people dream up to justify continuing the 10,000-pound threshold? Lobbyists focus on two points. First, they argue that it would be too much trouble and too expensive for companies to report all toxic chemicals. Second, they claim that local agencies would be unable to deal with all the information that would be delivered to them in any case. Both arguments can be easily dismissed.

Congress, in passing the original legislation, explicitly stated that the costs and troubles to industry were not to be a factor in deciding reporting thresholds. So much for the first point. As for the second, with dramatic increases in computer technology there is no evidence that local officials would be unprepared to handle a large volume of information. Indeed, the International Association of Fire Fighters has been strenuously fighting to ensure that the 10,000-pound threshold not be made permanent. And where lower thresholds have been implemented by local law, everything seems to be working smoothly. The fire department in Lancaster County, Ohio, for example, obtains information on all hazardous chemicals present in quantities greater than one pound. The department has reports on approximately five hundred dangerous chemicals present in a single Ralston Purina plant and feels prepared to deal with any emergency that might arise in that plant. If the 10,000-pound threshold were in effect in Lancaster, that same fire department would have information on only six of the chemicals. So much for industry's second point.

The final piece of critical safety information which industry has been categorically unwilling to make public concerns internal self-studies. Fred Millar, director of the Toxics Project at the national environmental group Friends of the Earth, says that even those companies that have spent millions of dollars performing detailed hazard assessment studies have steadfastly refused to share any of their worst-case scenarios. In a phone conversation, Millar claimed that he has been unable to find a single company willing to make such information public; one company offered its study to an LEPC on the condition that the LEPC keep the information confidential. The reason for this secrecy, Millar states, is that the companies believe that "there will be panic in the streets when people find out what they know about these risks." Although it might not be unreasonable to argue that it is wise to avoid public panic, it is also perfectly true that people have a right to be informed of the risks to which they are being subjected.

Furthermore, lawmakers can make rational policy only when meaningful information is available to them and their constituents. The chemical industry has repeatedly taken the stance that just such information should not be disclosed, partly claiming that the public, and even its elected policymakers, are too unsophisticated to deal with that sort of technical data. In reality, however, lack of disclosure fuels existing fears.

One of the things that people are afraid of is pesticide use, so it is not surprising that a good deal of legislation has focused on the safety of the active ingredients in pesticides. Aside from the fact that even with such legislative attention the vast majority of active ingredients have not been subjected to a full array of health and safety checks, active ingredients represent a vanishingly small percentage of the concoctions sprayed on our fruits and vegetables and dispersed throughout our environment. They typically constitute no more, and often much less, than 1 percent of the product. The remaining 99 percent of the solution is benignly referred to as inert.

Those inerts, however, may be nothing of the sort. In many cases, they are killers and carcinogens in their own right. Strangely, it is the pesticide manufacturers themselves who define what is and what is not inert, and the EPA requires them to do so by a process of elimination. Rather than being forced to assess the properties of each particular component, manufacturers are required to designate which ingredients they feel "prevent, destroy, repel, or mitigate any pest." The rest, regardless of actual composition or effect, are grouped together under the heading "inactive." This bizarre process ensures that the active ingredient in one company's brew is occasionally the inactive element in another's. Complicating matters further, in most cases, manufacturers are not even required to list their inactive ingredients. These data are considered to be proprietary information and, as such, are protected by law. Given the present situation, the National Coalition Against the Misuse of Pesticides has made the reasonable suggestion that "inactive" be changed to "secret" on all labels.

Yet even if manufacturers were forced to disclose fully the composition of their pesticides, we would not be much better off than we currently are. The simple fact is that the EPA has remarkably little information about the hazards associated with the vast majority of inerts. Of the approximately 1,200 chemicals known to be used as inerts, the EPA lists only 300 as safe. An additional 100 are listed as "potentially toxic," and the EPA is unable even to categorize the remaining 800 because the necessary tests have not been performed.

Even if the hazards of each inert were known, the problem would be far from solved. Inactive constituents together, or in combination with active elements, often form solutions with entirely unexpected properties, unlike those of the inerts or of the actives alone. The active ingredient in the insecticide malathion, for example, penetrates skin much faster when it is mixed with an often used inactive ingredient, xylene. Kerry Leifer, a chemist in the EPA's registration division, acknowledged, "Maybe by themselves [the inerts] are not toxic, but the mixture could be. But to take into account all the possible permutations is beyond the scope of our program."

Citizens, then, should demand that the scope be broadened. Instead of requiring tests and licensing of specific components of pesticides, the government should require that the health effects of entire products be examined. A company wanting to produce a pesticide should need to prove to the government that that pesticide is safe— not just the active ingredients, not just the inactive ingredients, not even some combination of the two, but the actual product that is being marketed. With such a rational policy in place, consumers would be markedly safer than they are today.

Research conducted by Susan Fisher, environmental toxicologist at Ohio State University, demonstrated yet another related problem with the testing of pesticides. Her studies have shown that pesticides behave differently under a variety of environmental conditions. "The EPA and industry view the environment as a generic one, one which has little semblance to reality," cautions Dr. Fisher. Real pesticides in the real world are used under widely varying conditions. Many chemicals behave very differently in an acidic environment than they do in a neutral one, in a hot rather than a cold one, or in a wet rather than a dry one. To complicate matters even more, diverse environmental conditions interact with one another to produce even stranger chemical behavior. Fisher found that the toxicity of some pesticides can vary more than a hundred-fold depending on the environmental conditions present at the time of application. Some compounds that should be broken down to harmless by-products within days of application have been found to remain active for as much as a year later. Pesticides typically show both greater toxicity and persistence under hot and dry conditions, suggesting that the risk of serious pesticide contamination is much greater during droughts or in ordinarily arid environments.

Even home garden products pose serious risk. Consider the situation with carbaryl, often sold under the trade same Sevin. Its manufacturers claim that foodstuffs treated with their product can be harvested and safely eaten a few days after application. Fisher views that claim with a good deal of skepticism because carbaryl is broken down by microbial action, which in turn is directly dependent on a whole suite of environmental characters. Harvest too early for your particular environment and, along with your vegetables, you might well ingest a mutagenic chemical known to destroy kidney function. An informed public must demand that appropriate testing be performed before manufacturers are permitted to market their products.

The idea of placing the burden of proof for the safety of a product or a chemical on the manufacturer rather than on the government makes obvious sense. As Garrett Hardin, prolific author and professor emeritus of human ecology at the University of California, Santa Barbara, said in his book *Living within Limits: Ecology, Economics, and Population Taboos:*

> Focusing on criminal acts, ancient Anglo-Saxon law had decreed that the default position of the law should be "innocent until proven guilty." For the criminal law this is undoubtedly the best assumption. But when it comes to laws that govern the everyday activities of citizens living in an ever more crowded world, the assumption is perilous. Chemists have synthesized more than a million compounds, and a wealth of experience indicates that the effects on human beings of most of the compounds is [*sic*] bad. Society would soon be bankrupted if it had to prove, in courts of law, the harmful effects of every one of those compounds, while determining the threshold concentration of each at which harm is first observed. The most rational policy is to put the burden of proof on the entrepreneur who is hoping to make a profit from introducing one more compound into the human environment. The cost of proving harmlessness then becomes one of the costs of a profit-oriented business—as it should be. (201–2)

Even though federal legislation ostensibly has been moving us in this direction, we still have an enormous distance to travel. Far too many chemicals whose safety has not been appropriately certified are currently in widespread use. It surely is not too much to ask that public health be placed before private profit. Such a request would not make us either antichemical or chemophobic. It simply means that we believe in moderation when it comes to using potentially lethal products and that we believe that we should have enough information to be able to formulate rational public policy. As many of the examples described above indicate, the chemical industry, often with the acquiescence of the government, is refusing to disclose far too much of that information.

A knowledgeable citizenry needs to work with governmental agencies to demand more openness on behalf of industry and to reduce dramatically our reliance on chemicals. Reducing chemical usage should not be seen as an unachievable, idealistic desire. The U.S. National Academy of Sciences released a major report in 1991 demonstrating that widespread pesticide use is not in the public interest and calling for major reductions in the use of agricultural chemicals. Their report seems to have already been heeded in some corners: a *Science* report indicates that pesticide use is on the decline while crop yield is rising. The surprising thing is that the *Science* results reflect changing agricultural practices in the Third World rather than in developed nations. The improvements have been very impressive. A shift from pesticide-based agriculture to an integrated pest management philosophy in the rice fields of Indonesia, for example, has yielded a 10 percent increase in rice, a huge decrease in capital outlay for chemicals, and an untold diminution of pesticide-related illness. In Bangladesh, farmers using integrated pest management techniques reportedly spent 75 percent less money on pesticides while harvesting a crop almost 14 percent larger than did those farmers using pesticide-intensive methods.

That's the good news. The bad news is that the developed world remains stubbornly addicted to the use of unnecessary toxins. David Pimentel, one of the world's leading agricultural experts and a biologist at Cornell University, estimates that more than five hundred species of insects are now resistant to pesticides and that the amount of crops "lost to insects has almost doubled during the last 40 years despite a more than tenfold increase in the amount and toxicity of insecticide."

Nonetheless, First World countries seem completely unwilling to act on this knowledge. The Indonesian situation demonstrates not only that dramatic governmental action is necessary to effect major changes but also that dramatic governmental action is possible. In 1986, with rice crops decimated by a voracious insect known as the brown planthopper (in that year alone, the brown planthopper ate enough rice to feed three million people), the Indonesian government decided that extraordinary action was needed. Officials began by listening to people like Peter Kenmore, manager of the United Nations Food and Agriculture Organization's Inter-country Program for Integrated Pest Control in Rice in South and Southeast Asia. Kenmore was able to explain that most pesticides, in addition to killing insects acting as pests, also kill the beneficial, predatory insects keeping the pests under control. In essence, increased levels of pesticide usage had greatly exacerbated the brown planthopper problem. Kenmore's graphic use of metaphor helped explain the situation: "Trying to control population outbreaks with insecticides is like pouring kerosene on a house fire" (quoted in *Science,* 29 May 1992, 1272).

The Indonesian government's willingness to listen was followed by two courageous actions. First, government subsidies for pesticide distribution were eliminated, and the money saved was earmarked for integrated pest management training sessions at the local level. Second, fifty-seven of the sixty-three pesticides used on rice were banned. In addition to increased harvests, the reduction of pesticide usage has improved local water supplies, a point of particular importance because a 1990 study supported by the Rockefeller Foundation found that the pesticides commonly used in rice paddies can lead to a host of human maladies ranging from skin ailments to heart disorders.

Similarly courageous action is desperately needed in the United States. Currently, approximately 750 million pounds of pesticides are broadcast annually across our country. More than three decades ago in *Silent Spring,* Rachel Carson alerted us to the ecological consequences of this environmental assault. More recently, the National Academy of Sciences concluded that our abuse of pesticides is responsible for approximately twenty thousand additional cases of cancer in the U.S. population annually because "the average consumer is exposed to pesticide residues . . . in nearly every food."

Three critical questions come to mind. Is the developed world willing to learn from the successful example of the Third World? Do we have the courage to take the advice of our own leading scientists? Are we willing to suffer the consequences of answering either of the first two questions incorrectly?

Chemophobia indeed. All of us would be much better off if we developed a healthy respect for the noxious chemicals that pervade our lives. If we could develop appropriate protocols for handling such chemicals and, as a society, work diligently to reduce their use, we could have a significant impact on our own health and well-being.

Chapter Eight

ॐ

You Are
What
You Eat

Given the reaction of the federal government, you'd think that a handful of supermarket chains had committed a serious crime in September 1989. Immediately after the chains announced their actions, governmental officials scrambled to hold news conferences to censure the actions of the supermarket owners. What dastardly thing did the supermarkets do to spark such a flurry of action? They agreed to ask food suppliers to disclose which pesticides were used in growing fruits and vegetables and further agreed that they would share such information with customers. Additionally, they said that they hoped such action would encourage growers to phase out the use of sixty-four pesticides that the Environmental Protection Agency lists as the greatest potential hazards to human health. So far so good, right?

Surely the chains must have done something else to provoke the EPA to say it was "troubled" by their actions, or for the National Agricultural Chemicals Association, the mouthpiece for the $4 billion per year domestic pesticide industry, to call the action irresponsible. Unbelievably, however, that's all the supermarket chains did. Reflecting the interests of their customers, they simply exercised a bit of the consumer power that is the mainstay of a free-market economy. And they did it just days after the National Academy of Sciences released a report demonstrating that widespread pesticide use is not in the public interest and calling for major reductions in the use of agricultural chemicals.

Given the amount of money at stake for the pesticide industry, you can understand its partisan position and almost keep from laughing when its spokesperson claimed that the action was taken by "narrowly focused special-interest groups." But what really is astounding is that the EPA took an equally negative stance. The logic advanced by the EPA was, to say the least, odd. In a statement released in response to the supermarkets' action, the EPA claimed that the agreement "puts the onus of pesticide regulation on grocery retailers, who are not and should not be responsible for such decisions."

Of course grocers and shopkeepers should not be in the business of regulating pesticides. That is, after all, one of the reasons why the EPA was established. But the EPA has consistently shirked its duty in this regard. Ignoring the fact that the vast majority of chemicals that it supposedly regulates have been insufficiently tested for toxic effects, the agency's record on dealing with those chemicals that it knows to be deadly is absolutely appalling.

Consider the case of dinoseb, a carcinogenic and mutagenic herbicide commonly used for weed control on a variety of crops including lentils, chickpeas, blackberries, and boysenberries. In 1986, Michael McDavit of the EPA's pesticide office said, "We had enough data. We had to act," as he announced a ban on the toxic herbicide. Less than two years later, however, under severe pressure from the pesticide industry and without any new information suggesting that the chemical was even remotely safe, the ban was altered to allow remaining stocks of the poison to be used commercially. The closest the EPA came to protecting the public was its decision that women of childbearing age were not to apply the herbicide. As we saw in the previous chapter, such a stance is ill conceived at best.

Or consider the case of captan. As early as the late 1970s, the EPA was concerned about the carcinogenic effects of this commonly used fungicide. In 1980, because of these concerns, it began a special review. By 1985, the data were clear enough for the EPA to propose a plan to ban the chemical. Note that they didn't ban the pesticide; they merely proposed to ban it in the future. Then, in February 1989, four years after thinking about proposing a ban, the EPA acted on captan—it approved its continued use on the twenty-four fruits and vegetables to which it had been most commonly applied. Captan residues have been abundantly found on apples, cherries, grapes, peaches, strawberries, and watermelon.

Given this record, it is hard to dispute the supermarkets when they claim to be taking action because the EPA has been too slow to remove hazardous pesticides from the market. Rather than criticizing the supermarkets, officials at the EPA would have been wiser to rejoice that the business world was making an effort to reduce the use of products about which the EPA had grave doubts. Informed consumers, acting together, can exert enormous influence over the foods that are offered for sale. This was perhaps seen most clearly in the response to the Alar scare a number of years ago. Daminozide, or Alar, as it is popularly known, was voluntarily removed from domestic distribution in 1989 because of consumer concerns well before the EPA got around to taking formal action.

All facets of the Alar episode are instructive. At the height of the scare, in a display of style rather than substance and of image rather than information, Kenneth W. Kizer, director of the California Department of Health Services, publicly ate an apple at a news conference. The spectacle immediately brought to mind similar antics occurring over the years by pesticide manufacturers and proponents of nuclear power. Who can forget their offers years ago to down glasses of DDT, malathion, and plutonium?

It is a shame that governmental officials had to stoop to such a level in their attempt to educate the public about health concerns. The public deserves to have its public officials address the substance of the issue rather than participate in publicity stunts. The issue was not the safety of any one apple. Even the Natural Resources Defense Council (NRDC), the environmental group that raised public awareness about Alar, stated quite clearly that it did not fear any single apple but rather the effects of accumulating pesticide residues over a lifetime. NRDC scientists claimed that "more than half of the lifetime risk of developing cancer from exposure to carcinogenic pesticides used on fruit is typically incurred by the time a child reaches age six." Additionally, voluminous evidence has demonstrated that children are much more vulnerable to carcinogenic substances than are adults. To imply otherwise, as Kizer did, is dishonest.

In one sense, the NRDC decision to promote its study, "Intolerable Risk: Pesticides in Our Children's Food," so aggressively was a high-risk strategy. Lifelong eating habits are formed early and if the report encouraged parents to alter children's diets away from fruits and vegetables to processed sweets and red meats, health problems, including increased rates of colon cancer, would invariably result. Yet given the findings of the NRDC scientists, it would have been irresponsible not to act. The least controversial of the NRDC findings adequately demonstrates this point. To determine the health risks associated with any particular food, the Environmental Protection Agency has to estimate how much of that product the average individual consumes. The EPA estimates, for example, that apples constitute about 2.5 percent of the diet of the average American. The NRDC report pointed out, however, that children consume apples in vastly greater quantities than do adults. In fact, apples represent almost 13 percent of the average child's diet, or more than five times the percentage of that of the average adult.

The three-decade history of Alar since it was first registered for use graphically demonstrates the point made in the preceding chapter: instead of forcing companies to demonstrate that the chemicals they manufacture are safe, the government generally assumes that such substances are harmless. And once registered, the government finds it extraordinarily difficult to change its mind and ban chemicals that are subsequently shown to be dangerous. Although daminozide was first registered in 1963, was approved for use on apples in 1968, and was found, in five separate studies between 1973 and 1984, to be correlated with tumors of the lungs, kidneys, and blood vessels in laboratory animals, its manufacturer, Uniroyal Corporation, conducted no carcinogenicity testing until forced to do so in 1986 by the EPA.

It's not as if the EPA did not recognize the possibility of a danger from Alar. As early as 1980, the EPA informed Uniroyal that it might take steps to ban Alar. And on 1 February 1989, more than three weeks before the release of the NRDC report, acting EPA administrator John A. Moore said that "there is an inescapable and direct correlation between" the ingestion of Alar and "the development of life threatening tumors." Amazingly, having just made such a strong statement, Moore then announced that final regulatory action on Alar was being delayed for an additional eighteen months. NRDC lawyer and chemical expert Albert Meyerhoff pointed out at the time that the EPA "would never allow the chemical on the market now but lacks the power to take it off."

NRDC, with the help of strong statements from actress Meryl

Streep and an exposé by CBS's *60 Minutes,* convinced the public that the risks posed by Alar were not worth the benefits, mostly delayed ripening of sprayed apples. Consumer pressure forced growers to stop using the chemical and forced Uniroyal to halt domestic distribution. As soon as the product was pulled from the market, however, an organized backlash against the NRDC was started. Consider just two attacks published within weeks of NRDC's victory.

— Syndicated columnist William Safire wrote, "In the apple scare, minimal risks were inflated into imminent threats." Safire then went on to say that the NRDC "will continue to undermine the trust laymen have had in the scientists at the FDA, EPA and surgeon general's office."

— Elizabeth Whelan, executive director of the American Council on Science and Health, made her statement, cited in the previous chapter, about the country's need for a national psychiatrist to deal with its nosophobia. She claimed that the "most virulent episode of national nosophobia was triggered by the release last month of a report by the NRDC which opined that trace levels of agricultural chemicals cause cancer in kids."

A scientifically naive public might well buy the arguments advanced by those who, like Safire and Whelan, have the rare but wonderful ability to turn a phrase and create a striking image. The problem is that the content of technical scientific studies is rarely as accessible to the general public as are newspaper columns, nor is the language nearly as captivating. But language alone, no matter how captivating, cannot alter basic scientific facts found in technical research reports. A remarkable study jointly performed by researchers at Harvard University, McGill University, and the U.S. Department of Health and Human Services is an excellent case in point. Their highly technical paper was published in an issue of the scholarly journal *Risk Analysis* and has received virtually no public attention even though it called into question the basic assumptions of how the government estimates the risks posed by certain chemicals.

At issue is whether or not the government's estimates of risk are conservative. Virtually everyone recognizes that, with respect to risk assessment at least, conservatism is a virtue because, in this context, it means overestimating the risks posed by a chemical. Because experiments on humans obviously cannot be performed and because extrapolating from animal data is so uncertain, the actual hazards of chemicals are extraordinarily difficult to estimate. Given all of this uncertainty, the federal government has operated with the understanding that, from the standpoint of public health, it is far better to err on the side of safety—better to find risk where none exists than to miss it when it is actually there.

The *Risk Analysis* paper tested the accepted governmental models using available data on 309 chemicals to see if the models did indeed yield conservative results. Surprisingly, the scientists found that in a significant number of cases the models, rather than making conservative predictions, actually underestimated the risks. Coauthor Edmund Crouch of Harvard said, "We have no idea how big that underestimate is." Nor does this timely work allow us to predict when such underestimates occur. Coauthor John Bailar of McGill said, "We just do not have evidence that our risk assessments for chemical hazards are substantially conservative." But rather than burdening themselves with the type of analysis so painstakingly undertaken by the authors of the *Risk Analysis* paper, Safire, Whelan, and others ignored those findings and instead used colorful language to downplay risks that they were in no position to judge. Conservatives should be conservative when it matters most, lest someday they have to eat their own words.

The disinformation campaign that constituted the backlash against environmentalists in general and NRDC in particular has been quite successful. Now, years after Alar was removed from the domestic market, it is fair to say that the prevailing view among the general public is that such extreme action was unnecessary. This is a shame because the lesson that should have been learned was that an informed public has enough power to force industry to discontinue the use of nonessential and dangerous chemicals. Furthermore, what should be obvious from the present health of the apple industry is that profits (for others than the manufacturers of chemicals) are not dependent on overuse of chemicals.

Although many would have us believe that the Alar episode was overblown, reanalysis of the data indicates otherwise. Writing in *Science* magazine (7 Feb. 1992, 664) Adam M. Finkel, staff member at Resources for the Future, concluded that "the excess risk to many children was roughly 1 in 4000, or 250 times the 1 in 1 million standard generally regarded as *de minimis*." Increasing the cancer risk experienced by young children by this amount hardly seems like a conservative thing to do. Collectively, we should be proud that our actions reduced this unnecessary hazard. Indeed, a collection of eight individuals, including doctors, scientists, and a labor leader, trying to correct the misinformation promulgated about the Alar situation, wrote in the same issue of *Science:* "As long as cancer remains one of the leading causes of disease and death in our society, the prudent course is to reduce and avoid exposure to carcinogens, particularly those that are unnecessary. The removal of Alar without impacts on apple production was an important step toward the goal of decreasing unnecessary and avoidable exposure to carcinogens in the food chain" (664).

The courts have agreed that the actions taken were appropriate. Eleven Washington State apple growers filed a $200 million civil suit against NRDC and CBS News for spreading "false, misleading, and scientifically unreliable statements about red apples." U.S. district court judge William Neilsen of the Eastern District of Washington dismissed the case, saying that the NRDC report that started the furor, *Intolerable Risk: Pesticides in our Children's Food,* "is not a polemical tract preying on raw emotions and irrational fears" (quoted in *Amicus* [fall 1992], 7). Judge Neilsen went further and demonstrated that he was able to recognize what so many industry representatives refuse to acknowledge: "Serious consequences develop during the first several years of life, and what an adult might be able to safely tolerate has no bearing on the quantity or type of hazardous chemicals which a toddler can handle without adverse effects."

None of us should permit ourselves to be duped into thinking that the Alar situation was anything other than a major success. We should look to it as a confirmation of what can be accomplished while understanding that we still have quite a long way to go. Even the EPA has recognized that our almost indiscriminate use of pesticides is unsound. In February 1987 it ranked the presence of pesticides in foods as one of our most serious health and environmental concerns. What is so frustrating, and what makes the 1989 actions of the supermarket chains so important, is that we, as consumers, have so little readily available information about the extent of pesticide contamination of our fruits and vegetables.

Lawrie Mott, senior scientist at NRDC, and Karen Snyder, research associate at NRDC, analyzed the data in a wealth of technical reports to try to determine just how bad the situation is. Their findings, published by Sierra Club Books in a work entitled *Pesticide Alert: A Guide to Pesticides in Fruits and Vegetables,* warn us that we have little reason to be complacent about the progress we have made. They discovered, for example, that although DDT, dieldren, and chlordane were among the first (and virtually only) chemicals banned for use on agricultural products by the EPA, they still show up in disquieting amounts. When their book was published in 1987, DDT, dieldren, and chlordane were three of the five most frequently detected pesticides on potatoes, even though they had been banned years earlier.

Beyond those three banned substances, Mott and Snyder report that thirty-five additional chemicals are regularly found on fruits and vegetables. Carcinogenic chemicals such as acephate, chlorobenzilate, and parathion; mutagenic ones such as Captan, iprodione, and chlorothalonil; and such compounds as demeton, dimethoate, and cyhexatin, known to cause birth defects, are among the most commonly detected. Consumers should know that 63 percent of all strawberries and 55 percent of all peaches tested had pesticide residues on them. Similarly, they should know that only 1 percent of corn and bananas were contaminated. Alarmingly, pesticide residues were found on 48 percent of all fruits and vegetables sampled. Information of this sort might well help shape healthier diets as well as encourage growers to rethink their chemical usage.

As troubling as these figures are, the situation is even worse abroad. Although "only" 11 percent of the domestic cantaloupes tested were found to contain pesticide residues, a full 78 percent of those imported were contaminated. Similarly large differences were found between domestic and foreign bell peppers, cucumbers, and tomatoes. One could argue that the Food and Drug Administration (FDA), the agency responsible for regulating the importation of produce, should be doing a better job assaying pesticide contamination. Although the FDA probably should be doing a better job, that answer begs the real questions raised by tainted foreign produce.

As consumers we need to look at the patterns of preference we have fallen into, and we need to examine the legislative structure that permits, and perhaps it is not too strong to say encourages, unsafe pesticide residues on imported foods. There was a time when the produce sections of our supermarkets looked dramatically different from season to season. Now, with the possible exception of pumpkins in October and a few berries now and then, there is remarkably little difference between which fruits and vegetables we find in stock in January and which we find in June.

Although the lack of seasonality in produce makes shopping easier, it does not necessarily make us healthier. "Out of season" produce most often originates in foreign countries—what's out of season in Wisconsin is not necessarily out of season in Central America. Our constant desire for produce irrespective of local availability has a high environmental price tag. Just consider the inordinate amount of jet fuel used to fly bell peppers and tomatoes around the world. Furthermore, it is a fact that pesticide regulations in most developing countries, where the majority of our imported produce originates, are significantly more lax than are comparable domestic regulations, leading to higher levels of contamination. Even more significant, however, is the fact that U.S. chemical manufacturers have repeatedly found ways to circumvent the law, both implicit and explicit, and ship pesticides banned in this country abroad. Because it is the residues of these banned pesticides which are often found on imported products, those residues are often more dangerous than those found on domestic goods.

Consider the case of chlordane and Velsicol Chemical Corporation, its U.S. manufacturer. As was discussed in Chapter 3, the EPA banned all domestic agricultural use of chlordane in 1978—twenty-eight years after Arnold Lehman, then chief pharmacologist at the U.S. Food and Drug Administration, described chlordane as "one of the most toxic of insecticides—anyone handling it could be poisoned." Even with that long-term ban in place, controversy continues to swirl about the production of chlordane. Velsicol continues to reject the fact that there are dangers associated with chlordane. Instead, the company's bizarre press releases assert that "a large segment of the scientific community believes that chlordane and heptachlor [a constituent of chlordane] do not pose a human health threat." It is completely unclear who constitutes this supposed "large segment of the scientific community," for, as was pointed out in Chapter 3, chlordane has been shown to be a human carcinogen causing, among other problems, leukemia, aplastic anemia, convulsions, miscarriages, and birth defects. More important than the press releases produced by Velsicol is the chlordane produced by Velsicol. Our laws are so shortsighted that even though Velsicol cannot sell chlordane in the United States, the manufacture of unlimited tons of the chemical is still permitted. In 1991, Velsicol manufactured and exported over 1.1 million pounds of chlordane, or more than 3,000 pounds of the deadly substance each day.

Velsicol is able to get away with this because as a nation we have refused to ban the export of chemicals whose use we prohibit at home. Even if we were to take the crass position that is is none of our business what chemicals are used abroad, our self-interest should demand that an end be put to this ludicrous practice because, as we have seen, these dangerous chemicals are finding their way back home in imported food products. Troubling levels of chlordane have been found in imported beef, asparagus, mushrooms, fish, squash, chilies, and pickles. Beyond chlordane, estimates are that more than $1 billion worth of banned or never-registered pesticides are produced in the United States for export each year. And each year significant amounts of these deadly chemicals complete a morbid circle of poison by being imported back home as residues on a variety of foods.

Another cause for alarm is that many of these chemicals, like chlordane, are long lasting and highly mobile. Chlordane has been found in frighteningly high quantities in some surprising places. Fish, birds, tropical coral reefs, dolphins, whales, and even arctic polar bears far from any point of application have all been badly contaminated, as has human breast milk in Australia.

Long-lasting and mobile chemicals arising from sources other than pesticides are also causing significant contamination problems for foods. Some of these problems, like the ones associated with pesticides, are a direct result of our thoughtless environmental activities. It is doubly frustrating that as we become increasingly health conscious and shift our diets toward supposedly healthier foods, we are increasing our chemical burden. The pesticide residues on fruits and vegetables negate some of the health advantages of incorporating them into our diets. Similarly, as a nation we should benefit from the fact that we are now eating significantly less red meat and significantly more fish than we did just a few years ago. A growing set of scientific reports indicates that our rampant disregard for the environment has made this safer lifestyle a good deal more dangerous than it used to be.

The problem is that great numbers of the nation's freshwater fish are dangerously contaminated with mercury. This second wave of mercury poisoning is much more problematic that the first, which occurred in 1970. Then the bulk of the serious contamination was due to direct dumping of mercury by industries. Now its more pervasive dissemination results from the combustion of coal and the incineration of solid wastes. Previously, scientists thought that humans played a very small role in the amount of atmospheric mercury, but the latest studies have now shown that 75 percent of the mercury in the atmosphere is due to human activity and that the level has doubled since last century.

As with so many chemicals that find their way into the food chain, mercury accumulates as one moves up that chain of predators and prey. In aquatic systems the problem is compounded because fish do not seem to be adversely affected by mercury. As the fish continue to grow, they ingest more and more of the poison. People who eat heavily tainted fish, however, can be severely harmed. In humans, mercury attacks the central nervous system and leads to a series of maladies including, at low levels, the loss of sensation in the hands and around the mouth. With increasing dosage, unsteady gait, slurred speech, tunnel vision, loss of hearing, convulsions, dementia, and death follow. Developing fetuses are particularly sensitive; studies have shown that babies can be born with severe brain damage even when their mothers manifest no overt symptoms.

What is most unsettling is that fish populations in even highly remote and seemingly pristine lakes are heavily contaminated. Studies have shown that isolated lakes in Minnesota, Wisconsin, Michigan, and Sweden all contain fish with dangerously high mercury levels. Research has also demonstrated that 95 percent of the lakes tested in Ontario show acute levels of the toxin.

The one factor tying all of these studies together is that mercury levels in fish are greatest in lakes that are acidic. The reasons for this are complex but understandable. The amazing amount of mercury we spew into the atmosphere is carried on wind currents and migrates around the world. When it falls in acidic bodies of water, it is changed from elemental mercury into the highly toxic methylated form. Although scientists are unclear how the acid accomplishes the methylation, they are convinced that the combination of elemental mercury and acidic water is deadly.

What is completely clear is that the problem is big enough that large numbers of people are at risk. Roughly 40 percent of Americans currently eat enough freshwater fish that their yearly intake of mercury exceeds safe levels. A study of a Cree Indian population in northern Quebec found elevated blood mercury levels and an abnormally high incidence of neurological damage. After counseling, the Cree modified their traditional diet and reduced the amount of fish they ate. As a result, blood levels of mercury have been cut in half with a corresponding drop in new neurological problems. Finally, wildlife has also been affected. The reproductive failure common in eagle, mink, and otter populations surrounding the Great Lakes has been attributed to mercury toxicity.

The FDA is aware of the problem and refuses to allow interstate shipment of fish containing more than one part per million (ppm) of mercury. Minnesota, the state with the most stringent regulations, has set a limit of 0.16 ppm. To put these numbers in perspective, consider that bass taken from the Florida Everglades average 4.4 ppm. Not surprisingly, mercury poisoning has been established as the cause of death of loons and endangered panthers in that region.

As is so often the case, our activities have dramatically altered global patterns and have thrown a balanced system out of whack. But the lethal combination of acid precipitation and mercury poisoning can be ameliorated with a single remedy; we need immediately to cut back on our burning of coal and on waste incineration. Until then consumers should lobby for better information regarding environmental contaminants in food products and should limit the amount of freshwater fish eaten.

The question that immediately comes to mind is how much information should be provided to a largely scientifically illiterate public. In other words, is the public scientifically sophisticated enough to understand statistical arguments over possible increased risks arising from certain foods and contaminants? By overreacting to such information the public, in essence, victimizes itself. Such was the case with apples and Alar when people became frightened to eat single apples. A controversy within the neuroscience community is a wonderful example of the dilemmas facing everyone involved in public health issues. Neuroscientists agree that there is a class of amino acids that can function as excitotoxins. These particular building blocks of proteins, specifically glutamate and aspartate, stimulate neurons to high levels of activity when present in the brain. In excessive doses, these amino acids overstimulate nerve cells, leading to their death—hence the name excitotoxins. An overabundance of excitotoxins might well lead to damage of the hypothalamus and variants of both amyotropic lateral sclerosis (Lou Gehrig's disease) and Parkinson's disease.

There are numerous critical issues upon which the neuroscientists have, as of yet, been unable to agree. Given, for example, that both of these amino acids naturally occur in the brain, how much of a danger do they pose when they are used as food additives, glutamate typically in the form of monosodium glutamate (MSG) and aspartate as aspartame in artificial sweeteners? In addition, there appears to be mounting, but not yet conclusive, evidence that susceptibility to excitotoxins might well be genetically predetermined. Given the controversy within the scientific community, how much information should be made available to the public and how much of a leadership role should the government, specifically the FDA, play in establishing usage guidelines?

For the most part, the FDA has opted to stay uninvolved. Approximately a quarter century ago when neuroscientist John Olney discovered that infants were more susceptible to excitotoxic damage than were adults, he campaigned to get the FDA to ban the use of MSG in children's food. Although the FDA has refused to take such an action, baby food manufacturers have voluntarily agreed to discontinue its use. Nonetheless, Olney is still concerned that a small child might ingest a toxic amount of glutamate from prepared foods not specifically designed for children. He is worried, for example, that a six-ounce serving of a number of brands of instant soups might cause serious neurological damage in some children. Although Olney is by no means supported by all neuroscientists, neither is he alone in his concern. Indeed, at a national meeting of the Society for Neuroscience there was a full discussion of the possible dangers posed by excitotoxins. Unfortunately, despite an invitation, no members of the FDA were present.

By sitting out the action, the FDA is implying to consumers that there is no problem with excitotoxins used as food additives. If the FDA is correct for the majority of consumers, but not for children or for the small subgroup of adults possibly genetically predisposed to overreact to excitotoxins, what then? Perhaps the additives in question should be completely banned until all doubt is removed. Or perhaps, as some are recommending, the FDA should require warning labels on foods using additives containing excitotoxins. To be accurate and honest, however, those labels would have to convey the sense of the uncertainty present in the scientific community.

The balance between paternalism and undistilled information is a difficult one to achieve, and when vested interests are not kept at bay, dangerous consequences may result. In 1987, for example, when the Reagan administration was attempting to reverse a longstanding practice precluding food manufacturers from listing health claims on their labels, Otis Bowen, then secretary of the Department of Health and

Human Services, said, "We want to permit and, in fact, encourage science-based statements regarding the benefits which classes of foods can provide." The food industry was ecstatic over the proposed regulations, which conveniently stipulated that, although a government committee would be established to help develop appropriate messages, manufacturers would not have to receive clearance before displaying their health claims. Instead, the committee might be able to force errant advertisers to change their labels only after the fact. Basically the policy was one of self-regulation by industry and, in some critics' minds, would return us to times gone by when elixirs like Dr. Thomas's Electric Balm, Hamlin's Wizard Oil, Tom-Tom Herb Tonic, Modern Miracle Medicine, and Wa-Hoo Bitters were common and claimed to be universal panaceas, good "for man or beast." While they made for interesting reading and the vending of such creations made for fascinating entertainment, the products themselves served no medicinal purpose.

Fears that the public might well be misled by a proliferation of unsubstantiated or poorly substantiated claims are well founded. Even under the more stringent regulations currently in place, a whole new industry has been born as a result of misleading advertising. The calcium industry has increased dramatically in recent years with sales topping $150 million in 1986, because of the extravagant claims that advertisers have been allowed to make concerning the effects of dietary calcium on the prevention of the bone ailment osteoporosis. Labels of products from antacids to breakfast cereals are touting the amount of calcium supplied by their products, and advertisements clearly claim a link between increased intake of dietary calcium and reduced probability of debilitating bone loss. In addition to spawning a new industry, such advertising has influenced the eating habits of millions of Americans.

Are Americans healthier because of these changes? The data strongly suggest that the calcium craze, while working wonders for the calcium industry, has had no discernible positive effect on the health of adult consumers. (Everyone agrees that calcium intake during childhood and adolescence helps determine bone mass in adulthood.) Experimental studies performed at the Mayo Clinic and at Golstrup Hospital in Denmark have been unable to demonstrate any link between adult calcium intake and bone loss. These findings have not been surprising to the large number of scientists aware of the many population studies showing that those people naturally consuming the most calcium have no denser bones than those consuming the least. Richard Mazess of the University of Wisconsin has gone so far as to call calcium "the laetrile of osteoporosis."

Is increased intake of dietary calcium dangerous? Few studies have looked directly at this question. Although most researchers assume that no serious harm can come from high calcium levels, three problems might arise. First, high calcium intake might lead to kidney stones in susceptible people. Second, increased intake apparently can interact with vitamin D, thus interfering with normal activation of bone cells. Finally, the focus on calcium supplements diverts attention from what researchers agree are the best ways to slow the rate of bone loss in postmenopausal women: estrogen treatments and weight-bearing exercise. It is also interesting that in all of the media hype about the benefits of calcium, very little effort has been spent to teach people that the risk of osteoporosis is doubled by smoking and by as few as two alcoholic drinks per day.

Determining the best ways to educate consumers about health benefits and risks associated with certain foods, additives, and contaminants is extraordinarily difficult, yet it is clear that leaving that critical task to industry is a mistake. Consumers need to know that there are real health risks associated with pesticide and other chemical usage. Simply dismissing such concerns, as Elizabeth Whelan does, by referring to them as examples of "national nosophobia" serves no productive purpose. In addition, industry representatives need to be made aware of the huge difference between voluntary and involuntary risks. While it might be true, as some have claimed, that it seems odd for industry to worry about the fractional increased cancer burden caused by pesticide residues on the fruit eaten by someone opting to smoke two packs of cigarettes a day, such an argument entirely misses the point. It is one thing for individuals to voluntarily opt to inflict risks upon themselves but quite another to force even smaller risks on those same people without their consent. There are some added risks that individuals must bear because, by doing so, society as a whole is improved. Determining which risk falls into this special category is obviously the challenge for a democratic society.

As with so many other things, Rachel Carson, talking more specifically about pesticides, said it best. "It is not my contention that chemical insecticides must never be used. I do contend that we have put poisonous and biologically potent chemicals indiscriminately into the hands of persons largely or wholly ignorant of their potentials for harm. We have subjected enormous numbers of people to contact with these poisons, without their consent and often without their knowledge."

Chapter Nine

ॐ

Land of
Many
Abuses

Although roughly 80 percent of the earth's surface is covered by ocean, it is the remaining 20 percent that most of us call home. On this small offering of land, we build our homes, grow our food, harvest our timber, and extract our minerals. Over the millennia, our work and play have dramatically changed the face of the Water Planet.

These changes are surely not new. In biblical times, the hills of Lebanon were carpeted by magnificent cedars, prized throughout the ancient world. Solomon, when building his fabulous temple and palace, ordered "fourscore thousand hewers in the mountains" to harvest those wonderful trees. Today, Lebanon's hillsides are bare, and because of changes in microclimate directly related to the removal of these proud cedars, what was once forest is now largely barren and infertile.

With nearly two-thirds of the world's population dependent on firewood as its primary means of cooking, the demand for timber appears insatiable. As happened in Lebanon, when forests are cleared, the environment is dramatically altered, often with strikingly negative results. The term *desertification* has been coined to describe this process of converting productive areas into arid, desertlike stretches of land. Moderate desertification can occur with a decrease in agricultural productivity of 10 to 25 percent; severe desertification might result in a

productivity decrease of 50 percent or more. Damage this extreme usually results in the formation of massive gullies and sand dunes.

Deforestation is not the sole human cause of desertification.
Overgrazing of rangeland, cultivation of land with unsuitable terrain
or soils, and compaction of soils by machinery and cattle hooves, as
well as the impact of raindrops on the denuded ground, all impact negatively. The problem is by no means a small one, both in terms of land
mass and people whose lives are affected. United Nations statistics indicate that nearly 35 percent of the earth's land surface is classified as
arid or semiarid desert, with approximately 20 percent of the world's
population attempting to eke out an existence in these harsh locations.
And the problem has been growing dramatically. The situation has
been best described by G. Tyler Miller Jr. in the third edition of his
textbook *Environmental Science: Sustaining the Earth* (1991):

> The world's rural people affected by desertification rose from 57 mil-
> lion people in 1977 to 230 million in 1989. By the end of this century
> 350 million more people may be affected by desertification.
>
> It is estimated that about 810 million hectares (2 billion acres)—
> an area the size of Brazil and 12 times the size of Texas—have become
> desertified during the past 50 years. Each year an estimated 6 million
> hectares (15 million acres)—an area the size of West Virginia—of
> new desert are formed.

Although the people of many developing countries experience the most
pronounced hardships from the effects of desertification, it would be a
dangerous mistake for us to ignore its effects in our own backyard.
Consider the case of the high plains of the American Midwest. Illinois,
Missouri, and eastern Kansas, states whose lands used to support the
magnificent tall-grass prairie ecosystem, now are among some of the
most fertile agricultural regions in the world. But just a bit west of
there, across the ninety-eighth meridian around central Kansas, the situation changes dramatically. Here, the native grasses, instead of reaching six to ten feet tall like the native growth of its eastern neighbors,
consist of plants a third or half that size. The shift in agricultural patterns along this boundary are equally dramatic, a difference in large
part due to the pattern of rainfall—everything west of the ninety-
eighth meridian falls well within the rain shadow of the Rocky Mountains. Intensive farming and ranching are possible in these short-grass

prairie and high plains ecosystems only with massive subsidies of water and fertilizer. And even under the best of conditions the environmental effects of intensive agriculture in this marginal habitat are huge. Under less than ideal conditions, the negative effects reach biblical proportions; millions of pounds of topsoil take to the air, blackening the sky in frightening dust bowls. During one windstorm on 30 April 1930 approximately thirty-five tons of soil were deposited per square mile over eastern Nebraska, representing the erosion of much of the rich A-horizon (topsoil) of the upwind source region.

These are the biological facts. The demographic effects are just as striking. "The Plains are hemorrhaging people," explains Frank Popper in Anne Matthews' wonderful book *Where the Buffalo Roam: The Storm over the Revolutionary Plan to Restore America's Great Plains*. Economic hardship and ecological devastation, failing farms, a terribly eroded landscape, and dying small plains towns are forcing increasing numbers of people to evacuate much of the 139,000 square miles from the Dakotas to Texas and from Wyoming to Nebraska. With that exodus we are witness to a changing landscape. The American frontier—properly defined as any land including fewer than two people per square mile—which officially closed in 1890 has reopened with a vengeance. As of the 1990 census, 150 western counties must once again be considered frontier areas. Intellectually and emotionally we deal quite differently with a frontier that is being actively settled and one that is being created by a human retreat caused by environmental degradation.

Frank and Deborah Popper advocate the revolutionary idea that the proper thing to do would be to correct our environmental transgressions and return this gigantic rectangular piece of America to its natural condition. When viewed objectively, this audacious idea becomes much more than a misguided bit of nostalgia. The Poppers, he a land use expert and she a geographer, both at Rutgers University, have the data to back up their passionate arguments that creating a "Buffalo Commons" would simultaneously rejuvenate a long-lost natural ecosystem, improve environmental conditions worldwide, and generate a stable, tourist-based economy centered on the restored plant and animal life, all at less cost than it takes to continue the current practice of deluging the area with monumental federal farm subsidies.

Although such a wholesale change might be difficult, if not impossible, to achieve politically, thinking in such terms reflects a central ecological principle and might well offer productive alternatives in the nature of our discussions of habitat preservation. From an ecological perspective it is axiomatic that species, both flora and fauna, interact with one another in impressively complex fashions. If we want to maintain species diversity, indeed, if we want to prevent ecosystems from collapsing, we have to be aware of these complex interactions and work to preserve whole ecosystems rather than just threatened and endangered species. Fencing off and protecting a patch of prairie containing the federally threatened western prairie fringed orchid (*Platanthera praeclara*), for example, will do little to ensure the continued existence of this long-spurred, delicately scented wildflower if the long-tongued moths that serve as its pollinators are not protected as well. Protecting the moths means moving out from the small plot of prairie sod in which the orchid is rooted and working to preserve the web of species interactions which is the hallmark of healthy, flourishing ecosystems. Indeed, our goal must be much more than just re-creating a stagnant ecosystem. Healthy ecosystems, by their very nature, change and evolve, and it is this dynamic, natural state that we must work to preserve. A large part of the beauty and intrigue of biological systems lies in their capacity for change, something that can never be fully captured in any static museum display.

Preservation of entire ecosystems, unfortunately, has not been a part of America's conservation ethic. From the spectacular grandeur of Yellowstone National Park, created in 1872 as the nation's first such park, to the soaring peaks and brooding glaciers of Alaska's Wrangell–St. Elias National Park, created in 1980 as our most recent national park, the United States National Park Service has attempted to maintain and defend areas of dazzling scenic beauty or uncommon recreational potential while neglecting those parcels of land lacking extraordinary geographic features.

We have reason to be proud that most of America's scenic wonders have been afforded at least some modicum of protection. But consider the situation with the much discussed but not yet realized tall-grass prairie national park. Such a park, on the drawing board for decades, would preserve something other than those spectacular vistas that we have come to associate with our national parks. And because those vistas are lacking, the park has been much slower to take shape. A tall-grass prairie national park would maintain a rolling prairie of grasses and herbs, a prairie whose beauty is far more subtle but no less majestic than the peaks of Denali National Park in Alaska or the sculpted recesses of Canyon de Chelly National Monument in Arizona. Unlike other federal preserves, a tall-grass prairie park would include no single, unique feature of the American heartland but would save a portion of an ecosystem that originally stretched in an undulating sea from Ohio to Colorado, from Minnesota to Oklahoma. Such a park would preserve an ecosystem memorialized in "America, the Beautiful" but ignored by our political system. Once again, native bison and elk might wander freely on the great plains of North America.

Time is not on our side because ecosystem restoration is increasingly difficult as species are lost and as habitat is degraded. Additionally, as the native landscape cover of North America becomes increasingly developed and fragmented (and fragmented it has become—tall-grass prairies have been reduced by 98 percent, wetlands by 50 percent, and forests by 33 percent), we lose a sense of how these typical ecosystems operated. If we fail to protect the remaining portions of these habitats, we will forever lose the opportunity to study the biological processes responsible for the evolution of the intricate species interactions that constitute our continent's ecosystems. And we will lose a piece of our own history. Only standing in the middle of a tall-grass prairie, buffeted by the winds sweeping across the plain, can we obtain a glimpse into the lives of the early pioneers. No similar view can ever be realized in a museum or from a book.

Perhaps we will never flock to a tall-grass prairie as much as we do to Yellowstone or Yosemite. But examples of such typical ecosystems— which, by virtue of their former abundance, were subjected to a host of abuses—deserve a place in our national park system. Learning to admire and respect an environment that has historically been seen as typical, even uninteresting, challenges our basic conceptions of nature while encouraging us to seek beauty in subtlety, to value ecosystems for their own sake, and to recognize that our American heritage is best expressed in those landscapes most typical of our continent's primeval state.

Given federal legislation already on the books, we should have made significantly more progress than we have in terms of preserving ecosystems. The 1976 Forest Management Act states explicitly that national forests are to be managed in ways that guarantee "diversity of plant and animal communities." There are basic ecological reasons for doing exactly this. Introductory ecology courses stress the importance of interactions among species and repeatedly emphasize the synergy that occurs in ecosystems—that is, the whole is far greater than the sum of its constituent parts. As basic as these lessons are, not everyone is capable of learning them.

Three botanists from the University of Wisconsin, William Alverson, Stephen Solheim, and Donald Waller, have joined with the Sierra Club and the Wisconsin Audubon Council in filing a suit designed to force the Forest Service to rewrite its management plans for two Wisconsin forests to ensure that overall biodiversity will be maintained in keeping with the 1976 Forest Management Act. The scientists claim that the plan proposed for Nicolet and Chequamegon National Forests does not come close to meeting the act's stipulation. Alverson and Solheim should know. In the early 1980s, when the two men were starting graduate school and the plan for Nicolet and Chequamegon was being formulated, they were hired to survey the two forests for rare plants. Their documentation of approximately twenty rare botanical species, including a number of orchids, was largely ignored by the final plan. More strikingly, the plan omitted any discussion of management for biodiversity.

Waller, the most senior of the botanists filing suit, sounds like an instructor describing a student who refuses to study. "We were incredulous when we read the plans [and saw] that they had so abysmally misunderstood, misconstrued, or missed altogether all the information that was piling up out of ecology through the late 1970s and early 1980s" (*Science*, 18 Sept. 1992, 1618). One of the basic biological premises that had somehow escaped the notice of the Forest Service is that the shape of preserved land is critical. Many small preserves, for example, even if they have as much total area as one larger preserve, will typically be able to support many fewer species than will that larger area. One of the reasons for this stems from the fact that many environmental factors, such as temperature, humidity, and light penetration, are different along the margins of habitats than in the interior. Species that tolerate the internal conditions may be intolerant of the edge conditions and vice versa. The fragmentation caused by roads and clear cuts greatly increases the ratio of edges to interior spaces, and thus the array of species that can be supported in any given area declines rapidly as a function of fragmentation.

And it is the array of species present, rather than just the number of species, which is of greatest ecological and aesthetic importance. It is true, for example, that edge habitats often support more species than do interior habitats, but the particular species present often are quite different. Edge species are typically behavioral generalists that survive admirably under a wide range of conditions. These are rarely the threatened and endangered species whose environmental requirements are far more specialized. As any good bird watcher knows, edges of habitats are the places to go to see large numbers of common birds, whereas a visit to the interior of habitats rewards the naturalist with sightings of rarer species.

Alverson, Solheim, and Waller presented a revision of the proposed timber sales. Instead of fragmenting both forests, they proposed preserving a few 40,000- to 100,000-acre tracts as roadless, old-growth stands. These "diversity maintenance areas" would ensure the preservation of whole ecosystems and the species that compose them. Even though the botanists asserted, and many Forest Service personnel agreed, that this plan required no diminution of the total amount of logging in the forests, timber industry representatives adamantly opposed any modifications. The best explanation of the timber industry's inflexibility appears to be an objection, in principle, to allowing any environmental considerations whatsoever to play a role in defining the parameters of timber sales on federal lands.

The problem goes well beyond the timber industry, however. The concept of preservation seems anathema to some members of the Forest Service, the federal agency charged with stewardship of our national forests. Consider a letter written in 1992 by United States Forest Service regional forest supervisor Richard Wengart to Kentucky's congressional delegation. Wengart, informing members of Congress about grass-roots groups protesting alleged overlogging of Kentucky's Daniel Boone National Forest, cautioned that the environmentalists have a terribly dangerous environmental agenda "with the ultimate goal being preservation."

Wengart's was just one Forest Service attack on those attempting to minimize logging operations in Daniel Boone National Forest. Appalachia Science in the Public Interest (ASPI) found itself threatened with criminal prosecution for its role in protesting Forest Service policy. What was ASPI's alleged crime? Despite First Amendment guarantees of freedom of speech, the Forest Service accused ASPI of criticizing national environmental policy by displaying a caricature of a federal official. The federal official in question is Smokey Bear. (I know. When I grew up, Smokey was Smokey the Bear, but sometime over the past decade or so he officially lost his middle name and became Smokey Bear.) ASPI printed a leaflet decrying the logging practices in Daniel Boone National Forest, picturing Smokey in front of a field of stumps carrying a sign with the message "Only *you* can stop the Forest Service." The criminal prosecution threatened against ASPI was for "alteration of the Smokey Bear symbol and/or use of Smokey Bear for other than fire prevention purposes."

Similarly, the Americans for Ancient Forests (AAF) received a cease and desist warning from Forest Service chief F. Dale Robertson after running an advertisement protesting clear-cutting of old-growth forests. Their ad featured an ominous-looking Smokey brandishing a chainsaw above the caption "Say it ain't so, Smokey!" Bob Chlopak, director of AAF, was quoted in the January–February 1993 issue of *Sierra* magazine:

> It is not we, but the Forest Service, who has put a chainsaw in Smokey's hands. It is the Forest Service which has systematically encouraged the clearcutting of our last remaining ancient forests in California, Oregon, and Washington. If Smokey had standing, he just might seek to enjoin the Forest Service from these anti-preservation policies. What is needed is more public awareness and robust debate on these issues, not the threat of lawsuits designed to suppress such debate.

What is so frustrating about this situation is that the Forest Service is on the wrong side of two scientific issues. The first concerns the ecological importance of old-growth forests of the Pacific Northwest. Our concern for the preservation of these habitats should go well beyond our fondness for the spotted owl that has grabbed so much media attention of late and to the very integrity of the old-growth ecosystem. This concern is articulated in a summer 1992 open letter to Congress signed by 345 ecologists. The letter reads:

> As scientists deeply concerned about the worldwide loss of biological diversity, we are writing to express our support for efforts to protect the remaining old growth (ancient forests) of the Pacific Northwest.
>
> With respect for the diversity of conifer species and the size and longevity of individual trees, these forests are without equal. Whereas only two decades ago few people appreciated these old growth forests for their biological diversity, we now know they harbor over 200 spe-

cies of vertebrates and an uncounted number of plants, insects, arthropods, fungi, bacteria and other life. Together, they form an ecosystem whose richness and complexity we are just beginning to understand. These forests represent an irreplaceable, world-class resource; continued fragmentation of the ecosystem will only hasten a demise that may well prove irreversible.

Only about ten percent of the ancient forests of the Pacific Northwest remain today, mostly on federal lands where logging is permitted. Recent satellite images reveal an alarming rate of habitat loss and fragmentation. There are grave costs associated with the destruction of these unique forests, including the loss of valuable ecosystem services (such as clean water, spawning grounds for sensitive fish stocks, the absorption of carbon dioxide, etc.), the extinction of species and populations, and the loss of opportunities for scientific research and discovery.

We therefore urge Congress and the administration to permanently protect a sufficient amount and distribution of old growth forests and to mandate ecologically based management practices on other federal forest lands in order to provide a high likelihood of retaining a functional late successional/old growth forest ecosystem and associated species.

The satellite images referred to in the above letter, taken by NASA's Landsat-5 Thematic Mapper, show a distressingly clear picture of the magnitude of the habitat fragmentation that has already occurred and attest to the fact that the integrity of the Pacific Northwest forest ecosystem has already largely been destroyed.

In addition to ignoring the ecological importance of old-growth forest, there is a second scientific issue that has escaped the leaders of the Forest Service in their crusade to ensure that the only Smokey seen by the public is a Smokey saying, "Only you can prevent forest fires!" Simply put, forest fires are not all bad. Yes, recent years have been witness to a number of spectacular forest fires, some of which yielded equally spectacular and heroic efforts by firefighters. That the public has long been fascinated by such activity perhaps in part explains the great success of Norman MacLean's *Young Men and Fire*. Furthermore, forest fires can have devastating effects on humans, wildlife, and property. What seems largely lost in the media coverage of these blazes, however, is that fire is an utterly natural ingredient of most ecosystems and, when managed correctly, helps to preserve ecological diversity and integrity.

Our National Forest system was, in large part, created in response to an immense fire that swept through the forests of Wisconsin in October 1871, destroying the town of Peshtigo and killing some fourteen hundred people. Although this fire had not nearly as much news value as the great Chicago fire that occurred on the very same day, its devastation did spur Congress to act: the National Forest system was established soon thereafter. One goal of national forest management was the prevention of such fires.

When in the early 1900s the Division of Forests was changed to the U.S. Forest Service, President Theodore Roosevelt named the great conservationist Gifford Pinchot to be its first head. In this role, and later as dean of Yale University's School of Forestry, Pinchot proved a remarkably successful advocate of fire prevention and suppression. Ensuing decades brought Pinchot an extremely popular ally in Smokey Bear (née Smokey the Bear). Together they set federal policy and educated the public about the evils of forest fires. Their policy was simple: extinguish forest fires whenever and wherever possible. It wasn't until 15 or so years ago that the professionals in the National Forest Service, the National Park Service, and the timber industry began to allow many natural fires to burn and also to use fire as a management tool.

The shift away from aggressive fire fighting makes good sense from both management and biological perspectives. Aggressive fire fighting leads to large-scale buildup of flammable materials. Dead trees, branches, bushes, and assorted undergrowth all accumulate and serve as fuel. When this material is not reduced by small, periodic surface fires, the inevitable blaze will be far more extensive than would otherwise be the case. Under such conditions, surface fires expand and jump to the tree tops to become deadly crown fires.

From a biological perspective, there are at least three reasons why fires are beneficial in many forests. First, many of the tree species dominant in our western forests are relatively intolerant of shade. Seedlings of these species are unable to survive when adult branches begin to shade the ground. Over time, the diversity of tree species declines, and forests might become composed of only one or two shade-tolerant species. Of course, any reduction in tree species diversity also greatly limits the diversity of animal life. A related point is that highly shade-tolerant species often produce very dense forest. From an aesthetic perspective, such forests, which block out light and scenic views, are usually not very pleasing to hikers.

Second, many conifers depend on fires for the release of their seeds. Giant sequoia cones, for example, release seeds only after exposure to intense heat. Aggressive fire fighting greatly reduces opportunities for seedling establishment by these great forest inhabitants.

Third, periodic fires release many of the nutrients tied up in dead plant material. After a blaze, surviving trees are fertilized by the suddenly available foods.

Interestingly, by using fire as a management tool we are returning to the ways of some of the Native Americans. The Miwok Indians of Yosemite Valley, for whom acorns were a dietary staple, regularly burned cedars and pines in an attempt to promote the growth of oaks.

Needless to say, too much of a good thing can be ecologically disastrous. Extensive fires, fires that are increasingly likely to rage out of control when all previous blazes have been extinguished as quickly as possible, can lead to severe erosion problems. Animals have difficulty finding a refuge, and plant populations struggle, sometimes unsuccessfully, to recolonize a newly damaged area. Ecologists recognize that the healthiest situation is one in which the environment is composed of a mosaic of moderately sized burned and unburned areas. Once such a pattern is established, massive forest fires become much less likely. Recently burned areas serve quite effectively as natural fire breaks.

Fire is not the only natural process that humans have attempted to control, often to detrimental effect. The situation in coastal Louisiana, where the very land itself is disappearing out from under residents, is another striking example of the effects of overmanagement. The state of Louisiana is shrinking by approximately fifty square miles per year, and an area the size of Rhode Island is disappearing each generation. How has this come to pass?

Half the answer can be traced back to 1927, when massive flooding of the Mississippi River encouraged the Army Corps of Engineers to undertake the world's largest flood control project. It constructed a huge series of levees to keep the Mississippi within its banks. The Army Corps did its job extremely well, and now the river flows directly out to the Gulf of Mexico, no longer regularly overflowing its banks as it had done for eons. As we have seen so tragically during the summer of 1993, those levees cause severe problems during rare periods of very high water. Additionally, the very nature of the river environment has been altered by this construction. Now that the streamlined Mississippi no longer deposits sediments, the delta is steadily shrinking. Estimates indicate that 183 million tons of sediment per year (or 20,000 dump truck loads per hour) are being washed out to sea.

The second half of the answer involves the oil industry, which has riddled the area with about 8,200 miles of canals to service offshore oil rigs and pipelines. These channels hasten the process of erosion, thus adding to the rate at which Louisiana is washing out to sea.

As if erosion weren't trouble enough, the levees and canals are dramatically changing the ratio of salt water to fresh water in coastal marshes. The levees keep the Mississippi's fresh water from entering the marshes; the canals allow the salt water from the gulf to intrude. Statistics gathered by the National Audubon Society demonstrate that the problem is far from trivial—large numbers of freshwater marshes are dying. While 40 percent of the coastal wetlands of the continental United States (excluding Alaska) are in Louisiana, 80 percent of the country's wetland loss is occurring in that state.

Environmental and economic ramifications of the situation extend well beyond Louisiana. Because river sediments are rich in organic matter, healthy coastal wetlands are regularly and naturally fertilized and teem with animal and plant life. Louisiana's marshes have become the foundation for the state's $680 million yearly seafood industry. An abundant food supply is also the reason that 66 percent of the waterfowl in the Mississippi Flyway (the route migratory waterbirds use to navigate through the center of the country each year) overwinter in these marshes. As the habitat is destroyed, waterbird populations throughout the entire country will be decimated.

Fortunately, various environmental coalitions have begun to advocate changes in management plans. They have suggested, for example, that breaks be made in the levees to allow some flooding, thus allowing the deposition of sediments and replenishment of fresh water in the marshes. Restrictions on the construction and use of navigation canals as well as the restoration of barrier islands have also been urged. These solutions will work, although they are not going to be cheap.

But money is not the only issue. What we need to learn from the Louisiana situation is that, regardless of how technologically sophisticated we might become, we cannot expect to impose our will over nature without severe consequences. Instead, we need to develop an appreciation for natural processes, recognizing that it is as natural for rivers to overflow their banks as for forest fires to rage. Our sophistica-tion has given us the ability to step in and control these sorts of events even when intervention may not be in our long-term best interest. This may not be an easy lesson for us to swallow because it means that we have to acknowledge that we are a part of nature rather than above it.

As with forest fires, too much flooding, natural or not, can cause serious damage. Ironically, the same Army Corps of Engineers that has labored mightily to control the natural flooding patterns of numerous rivers has also exacerbated the problems associated with flooding by working to diminish the number of acres of wetlands offered protection from development. Wetlands, by acting as sponges for river overflow, often function as buffer zones during periods of high water, thus con-trolling flooding. The economic risks of wetland destruction become all too apparent when development projects are awash in floodwaters uncommon before the habitats were so dramatically altered. Addition-ally, as flooding increases, the amount of valuable topsoil that is washed away increases as well.

During the Bush administration, the Army Corps of Engineers was a strong advocate of the president's plan to exempt between forty and fifty million acres (constituting, by some counts, as much as half) of the remaining wetlands in the country from protection. The director of civil works for the corps, Major General Patrick J. Kelly, defended the exemption by claiming that the acreage involved only "minimally" met the criteria for wetlands. Minimally meeting criteria sounds, to me, suspiciously like being only a little bit pregnant. If the criteria are met, the land deserves federal protection, period.

Kelly had the nerve to go even further, however. He stated that the corps' decision to remove this land from federal protection "should not be construed as a weakening or a retreat" of the agency's commit-ment to preserve the country's wetlands. The only way to make sense of Kelly's statement is to assume that the corps had no such commit-ment in the first place. And there is ample reason to believe that this is so. In the past two hundred years, the United States has lost approxi-mately half of the wetlands present in the lower forty-eight states. The rate of destruction has been phenomenal. Wetlands have been con-verted to farmland, shopping centers, and housing developments at the astonishing rate of more than sixty acres per hour—every hour for two hundred years. California and Ohio lead the pack with more than 90 percent of their wetlands completely developed.

There are good ecological reasons to care about devastation of this magnitude. According to statistics provided by the National Wildlife Federation, wetlands provide habitat for fully half of all endangered plants and animals in this country. If wetland destruction continues, large numbers of additional species not yet endangered will be placed at risk. Wetland plants, often acting as filters, remove excess nitrogen and phosphorus from water and keep nearby lakes free of extensive and deadly algal blooms. Wetland habitats, by allowing water to purify itself as it percolates down to underground aquifers, help restock the nation's dwindling supply of drinking water.

Louisiana's wetlands notwithstanding, most previous conservation efforts have focused on coastal wetlands. Given this attention, the Army Corps of Engineers, along with lobbyists for developers, would have us believe that putting an additional sixty million acres of wetlands at risk is unimportant because those acres are mostly small and inland. What they cynically refuse to acknowledge is that the vast majority of our wetlands, almost 95 percent of them, are inland to begin with. And size is not necessarily an important indicator of the ecological significance of any particular parcel of land. Small prairie potholes, depressions created when large chunks of ice, buried beneath the soil during the last ice age, melted, providing nesting habitat for approximately 80 percent of all North American ducks. As these tiniest of wetlands have been destroyed, duck numbers have plummeted. The magnitude of the decline in redheads and canvasbacks is quite alarming.

In a visionary report released in November 1991, a blue-ribbon panel of scientists convened by the National Research Council (NRC) addressed the issue of wetland protection. Not satisfied with a mere call for the preservation of the nation's remaining wetlands, they advocated a "comprehensive and aggressive" restoration effort to reclaim previously destroyed habitats. In sharp juxtaposition to the Army Corps of Engineers' and George Bush's desire for a loss of forty to sixty million wetland acres, the NRC called for a restoration program yielding a ten-million-acre increase in such critical habitats by the year 2010.

The committee, perhaps with tongue in cheek, turned to the words of George Bush as its source of inspiration for its pro-active stance. "It is not enough merely to halt the damage we've done. Our natural heritage must be recovered and restored." John Cairns Jr., chair of the committee and an ecotoxicologist at Virginia Polytechnic Institute with a world-renowned reputation as a restoration ecologist, recognizes that his committee's recommendations will be expensive to implement. He points out, however, that the alternative is even worse. "Doing it now will be a lot more cost effective than postponing it even a decade or two. You only have to look at Eastern Europe and Russia to see the horror story of postponing environmental restoration" (*Science,* 10 Jan. 1992, 150).

Not all changes would be difficult to implement. During the massive flooding of 1993, the Mississippi River did what Congress refused to do—it reclaimed land that should not have been farmed in the first place. While dealing with the devastation brought about by that flooding, members of Congress seemed to have lost sight of the fact that there is a world of difference between wet lands and wetlands. As in previous years, Congress has been remarkably stingy about funding the Wetlands Reserve Program. Under this federal program, farmers with eligible land could offer to restore it to permanent wetland status in exchange for a government payment. Because of underfinancing of the program, only a fraction of the land that farmers wanted to enter has been accepted. In 1993 Congress allocated absolutely no money for the program, and in response to President Clinton's proposal to spend $370 million on it in 1994, Congress, after much wrangling, agreed to allocate only a mere $67 million. The irony of such a small allocation being made while the floodwaters of the Mississippi were still rising was not lost on Wendy Hoffman, a budget analyst with the Environmental Working Group. "Instead of paying farmers to restore wetlands, Congress will end up flooding the Mississippi Valley with disaster payments" (19 July 1993 EWG press release and phone conversation with the author).

Although few and far between, there have been some wetland victories. A coalition of environmentalists successfully fought off no fewer than four distinct attacks on land critical to the survival of a magnificent bird species. Although two wetlands used as breeding and feeding acreage were coveted by developers proposing a golf course to be associated with a new beach resort hotel, an industrial salt complex, and two separate roads, the government exercised the environmental foresight and courage necessary to refuse to issue permits for any of these projects.

Such a resounding victory is quite surprising given the government's wetland record. Why this particular turnaround? What provoked the federal government to at long last take a stand in favor of wildlife? The answer is actually quite simple. The magnificent bird saved is the greater flamingo, and the federal government having taken this virtually unprecedented action is that of Venezuela rather than the government of the United States.

There is a terribly important lesson to learn from this environmental victory. First, a bit of biology and a summary of the threats are in order. There is only a single population of greater flamingos in the southern Caribbean. These magisterial pink birds, often standing up to five feet tall and feeding with their heads upside down underwater, nest on the island of Bonaire, off the Venezuelan coast. During the wet season, breeding birds regularly make the ninety-mile flight to the Cuare Wildlife Refuge on the coast of the mainland. Teeming with shrimp and other crustaceans, the wetlands composing this refuge are the population's most important feeding ground. Because these wetlands are seasonal, the flamingos must look elsewhere for food when they dry out. Most often the birds move farther east to feed at Pirtu Lagoon. Cuare was threatened by the golf course and a road, and Pirtu was at risk from another road and the salt operation. Had any of these projects been permitted to degrade either of these wetlands, there is a good chance that the flamingo population would have gone the way of the dodo, the passenger pigeon, and Kirtland's warbler.

Now, the lesson. The seasonal wetlands of Cuare Wildlife Refuge, so central to the very existence of this species, are, to use the terminology of Marlin Fitzwater, Bush's press secretary, "mud puddles." (It was Fitzwater who said, in explaining how Bush was still standing behind his 1988 campaign promise of "no net loss of wetlands" while calling for a major redefinition of wetlands, "If you're from the school that says every mud puddle is a wetland, I don't think that makes common sense.") The Cuare example, much like the North American prairie potholes previously mentioned, dramatically teaches all of us that the importance of any particular wetland habitat cannot be determined simply by how wet it is, or for how long it remains wet.

Although the greater flamingos have been given a reprieve by the action of the Venezuelan government, another equally impressive tropical bird is in grave danger because of the activities of the United States government. Although it has received scant attention, the single tropical rain forest under the jurisdiction of the U.S. Forest Service is in profound jeopardy because of poor management. Puerto Rico's Caribbean National Forest, also know as El Yunque, is a magnificent remnant of the tropical rain forest that once covered much of the West Indies. Although at approximately 28,000 acres, El Yunque is the smallest of our national forests, it contains an impressive diversity of species divided into a number of distinct habitat types of remarkable range. At the lowest elevations, the forest consists of towering hardwoods reaching heights of more than 125 feet, and at its highest, the land is blanketed by dwarf forest. The entire area consists of 225 species of trees home to at least 1,200 species of insects. There are, as well, impressive numbers of mammals, reptiles, amphibians, and birds.

Among so many species, it is not surprising that some are extremely rare. El Yunque is the winter home of the endangered Arctic peregrine falcon and the permanent residence of the endangered Puerto Rican boa. But by far the scarcest inhabitant of the forest is the vanishingly rare Puerto Rican parrot. The most recent census indicated that there are, at most, twenty-eight of these beautiful green birds remaining in the wild, all in El Yunque.

With such an impressive array of wildlife, you would think that El Yunque would be a jewel to treasure at all costs. Indeed, the forest exists to the present day because it has enjoyed a long history of protection. Originally protected from early Spanish development by its designation as crown land, El Yunque passed to the United States after the Spanish-American War in 1898 and was quickly designated as a federal forest reserve by Teddy Roosevelt. As development destroyed the rain forest throughout the West Indies, El Yunque became unique. But inexplicably, the Forest Service and the Federal Highway Administration (FHWA) have been working overtime to put all of this at risk.

At issue is the desire of the FHWA to reconstruct a highway bisecting this biological wonderland. The road, originally built by the Civilian Conservation Corps during the depression, has been unusable for the past twenty years because of massive landslides. The slides were no surprise given El Yunque's steep grade, and even the Forest Service acknowledges that "any road construction and/or maintenance . . . will be risky and have a high rate of failure" (USDA memo, 20 July 1990). Nathaniel Lawrence, Natural Resources Defense Council attorney, lamented in a phone conversation the environmental havoc already caused by the road. "Acre upon acre of the forest is gone—it's just been swept away."

Given the potential for an environmental disaster, one would think that there must be compelling reasons for the FHWA to pursue this project so avidly. No such reasons have been forthcoming, however. Though the Forest Service has stated that they have not asked for the road to be reconstructed, though the U.S. Navy, which operates a communications facility on a mountaintop in El Yunque, claims the road "provides no advantage," though the president of a nearby resort hotel claims that he has been involved in no way with the project, the FHWA has remained completely silent in its determination to build the road. Given that far superior roads ring the forest and that any travel on a rebuilt road through the preserve would be incredibly slow going because of the steep grade and numerous hairpin turns, it is hard to dispute Lawrence's claim that this "is a road to nowhere."

Nonetheless, in the minds of those in power at the FHWA, this road has apparently taken on a life of its own. Currently environmentalists have halted the project by filing suit in federal district court (and winning two preliminary rounds). Their goal is to prevent the reconstruction of this road, the obvious habitat fragmentation that it would bring to this last vestige of West Indian tropical forest, and the ensuing demise of the Puerto Rican parrot.

Unfortunately, there seems to be an ecologically unpleasant association between roads and the destruction of tropical forests. While the road through El Yunque would endanger the smallest national forest in the United States, another road project has placed one of the largest protected tropical forests in Central America at risk. Developers have set their sights on Panama's 1.4-million-acre Darien National Park, declared a World Heritage Site and a Man and the Biosphere Reserve by the United Nations, home to the Ember-wounnan and Kuna indigenous people, as well as to a wealth of wildlife. Jaguars, ocelots, tapirs, monkeys, and giant anteaters populate the land, and endangered harpy and crested eagles soar overhead.

The proposal is to construct and pave a new section of the Pan-American Highway through the center of this biological treasure. Currently the highway extends from Alaska to the park's boundary, where it abruptly stops. The road continues beyond the borders of Darien National Park and the adjacent 178,000-acre Los Katios National Park in Colombia, reaching to the southern tip of South America. Closing the "Darien Gap" in the Pan-American Highway has long been a goal of those wishing to open the lush tropical forests of the two national parks to logging and mineral development. Only a series of successful lawsuits by environmentalists managed to halt construction in the mid-

1970s. The courts ruled that without extensive environmental studies and remediation, and without the eradication of widespread hoof-and-mouth disease in Colombia's cattle, the road project could not be completed.

Now, with the threat of the spread of hoof-and-mouth disease under control but with environmental arguments absolutely no better than before, the developers are back. Alfonso Araujo Cotes, Colombia's ambassador to Panama, has claimed that the United States will pay for two-thirds of the project's estimated $200 million cost. Juan Carlos Navarro, director of Panama's Association for the Conservation of Nature, fears that "opening a new road in Panama today is synonymous with deforestation, the destruction of nature and human misery." He points out that the consequences of the completion of the last stretch of the highway have "been ecologically devastating, with anarchic, uncontrolled deforestation, loss of soils, and the collective impoverishment of people who slashed and burned the forest to create fields" (14 Aug. 1992 press release from the Tropical Conservation News Bureau).

Profits and environmentalism need not be in conflict with each other, but they surely will be when developers lose sight of basic ecological principles. Consider the message being sent to students in Alberta, Canada, by a school curriculum, promoted by forestry companies, which states: "Clearcutting is the most effective way of foresting. . . . It is less expensive to reforest and other methods leave trees exposed to the wind and cold." Ignored is any discussion of the very serious ecological consequences of creating single-aged, single-species forests. Such forests are extremely susceptible to the spread of disease and lack the diversity necessary to support a complex, integrated web of animals and plants. The curriculum goes on to state, "Profit, after all, is the most important reason for our forests." There is nothing in this lesson plan about ecological, aesthetic, recreational, or moral imperatives associated with land preservation. A shortsighted, profit-oriented perspective on forests of a different, but no less insidious and significant, sort has reared its ugly head and is threatening to do grave damage to many pristine habitats.

The problem has been impressively explained by Robert K. Anderberg. Although you have probably never heard of him, he is a lawyer and writer worthy of praise. In a short *Amicus* article published by the Natural Resources Defense Council, he has done what so many before him have been unable to do—he has, in a simple and straight-

forward style, explained to the uninitiated, and perhaps even the uninterested, how corporate takeovers and increasingly frequent leveraged
buyouts work. In the process he has highlighted an impending environmental disaster that is a direct consequence of such actions. Corporate
raiders, it seems, have discovered the timber companies of the Northeast. The scheme, as Anderberg explains it, is so simple and so profitable that millions of acres of forest in New England are suddenly in
jeopardy. The basis of a takeover bid is the fact that the price of a company's stock is determined by the company's short-term profits rather
than by the actual value of the company's total assets. Timber companies in particular turn rather modest short-term profits, although they
typically have significant assets tied up in long-term investments—forests that are not to be cut for decades.

Consider the case of Diamond International (DI), taken over by
corporate raider Sir James Goldsmith in 1982. When Goldsmith took
control of DI, its stock was worth approximately $240 million, although
its total assets were actually worth about $315 million. Had he sold off
all of the assets, he could have turned an immediate profit of $75 mil-
lion, or 31 percent. Because Goldsmith performed a leveraged buyout—
he borrowed most of the money used to buy DI stock instead of using
his own—he was actually able to turn a profit of approximately 200
percent within two years. The profit arose because he was willing to
dismember the company; he sold paper mills, equipment, the corporate headquarters, and, most significant from an environmental perspective, more than one million acres of forests. The sale of all of this
land in and of itself is not something to be concerned about; timber
companies have sold land to one another for at least a century. What
really is of concern is that for the first time, huge tracts of forest land
in the Northeast were offered for private development. The consequences of such development will be striking.

Federal and state governments own only about 1.4 million acres
of forests in New England. Timber companies hold nearly 10 million
acres, more than seven times that amount, and much of the land owned
by the companies has been made available for public recreational use.
Given that most of this land adjoins national and state forests and
parks, it quite effectively serves as an environmental buffer zone between protected and unprotected habitats. As these acres are transferred from timber companies to condominium developers, opportunities for public use dwindle dramatically, as does the likelihood that

viable natural populations of animals and plants will survive. Environmentally, the effects of this sort of leveraged buyout might well be as
devastating as the effects of acid rain.

Although the northeastern states realize the consequences of this
massive land transfer, they have, so far, been virtually powerless to act.
New York State, for example, was soundly outbid by a Georgia-based
developer and land speculator for ninety-six thousand acres of pristine
habitat amid the Adirondack region. Similarly, a New Hampshire mobile home and condominium developer outbid the states of Vermont
and New Hampshire for ninety thousand acres of magnificent forest.
Even the timber companies themselves are powerless to help. Although
most of the remaining companies would prefer to acquire much of
this land for future harvesting, they are worried that to do so would
make them even riper for an unfriendly takeover and subsequent dis-
memberment.

Significant threats to ecosystem integrity from the transfer of
public lands to private control also arise because of the structure of
legislation enacted late last century during the presidency of Ulysses S.
Grant. Although his less-than-successful administration is perhaps best
remembered for numerous political scandals, it has left us with a piece
of federal legislation whose endurance equals its folly. The General
Mining Law of 1872 has been significantly modified only once, in 1920,
and is responsible for massive environmental problems as well as a tradition of supplying huge federal subsidies to the international mining
industry.

We should be fair to Grant and the federal legislators of his time,
however. Conditions and priorities were different a century and a quarter ago, and, in the context of the times, a law that encouraged both
mineral development and movement into the sparsely populated West
was easily perceived as a good thing. Based on the successful Home-
stead Act, the General Mining Law of 1872 allowed entrepreneurs to
file mining claims on virtually any federal land. These no-cost claims
transferred the rights to most hard-rock minerals such as gold, ura-
nium, copper, and silver from the government to the claim holder, who
was then free to extract any and all minerals desired without paying
any fees to the government. The law went even further. If the claim
holder could demonstrate that minerals were present in a quantity large
enough for any "prudent man" to turn a profit, the claim holder could
"patent" (or purchase) the land at a reduced rate of either $2.50 or $5.00

per acre, depending on the minerals present. Once patented, the land became private property and was no longer subject to any federal restrictions on its use. That the General Mining Law of 1872 is still on the books is reason for indictment not of the Grant administration but of our modern lawmakers. Because of the endurance of this law, massive amounts of federal lands have been converted to private ownership, the federal government has lost huge sums of money, and environmental devastation is rampant.

Although the law dictates that recoverable quantities of minerals must be present for a claim to be patented, it does not specify that those minerals must actually be mined. Once patented, therefore, the land can be used for any purpose. Land patented for $5.00 per acre adjacent to the Keystone, Colorado, ski area was sold by a mining company for condominiums for prices exceeding $50,000 per acre. Similarly, a seven-hundred-acre parcel of land in the Oregon Dunes National Recreation Area was patented for $2.50 per acre in 1989 by a mining company that then offered to sell it to the public at the market price of more than $17,000 per acre. An area the size of Connecticut has been transferred from federal to private ownership through the use of the patenting process, with no financial gain to the government.

Additionally, approximately $4 billion worth of minerals are removed annually from federal lands without the government receiving a penny in royalties or fees. Moreover, many of the mining operations are environmental disasters. Estimates suggest that more than ten thousand miles of streams have already been degraded by mining operations that produce runoff so acidic that it makes acid rain look like baking soda. Although federal environmental protection laws can be used to regulate mining operations on federal lands, they are rarely invoked because of the fear that mining companies will simply patent their claims and remove the land from public ownership.

And the situation is getting worse. Although Congress has been regularly discussing amending the General Mining Law of 1872, it has not yet been able to garner enough votes to pass any new legislation. The mining industry is, however, getting worried, and mining companies have responded by dramatically increasing the number of patent applications filed. Technological advances have also exacerbated the situation. Using a new "heap-leach" process that discharges such toxic compounds as cyanide, arsenic, and mercury into the environment, a "prudent man" can now expect to make a profit mining minuscule amounts of gold. If one ignores environmental effects, it can now be profitable to extract one dental filling's worth of gold from one ton of rock.

Given these developments, a moratorium on new patents is essential until the entire law is revised. That revision needs to include three essential parts. First, federal land managers must be given some discretion over when to approve mining claims. Although mining is clearly incompatible with many recreational uses and can destroy many natural scenic treasures, not to mention the deleterious effect it can have on wildlife, the 1872 law provides no such discretion, stating that no consideration can be given to any value other than mining. Second, environmental controls must be enforced on all mining operations. Strict regulations must ensure not only that mines do not foul the environment but also that mine sites are reclaimed after operations are finished. Finally, property reforms are needed. The patenting of federal lands should be abolished completely and replaced with a rental fee, with the federal government receiving royalties on all minerals extracted. A 12.5 percent royalty, the same amount the government charges for its outer continental shelf gas leases, would not be unreasonable. Such a fee would generate approximately $500 million per year, a significant portion of which should go into a fund to help offset the costs of cleaning up previously ravaged sites.

Mining and logging interests are not the only commercial undertakings that impose a significant negative effect on public lands and natural ecosystems. The cattle grazing industry has also been causing a number of serious environmental problems. Two situations are worthy of elaboration, both involving poor environmental decisions on the part of federal agencies charged with protecting the environment. The first concerns grazing in wilderness areas. Congress originally established federally protected wilderness areas with the thought that it was important, for both humans and wildlife, to enjoy a haven free from the modern world. In these habitats, wildlife might thrive and humans might appreciate the unspoiled beauties of nature. To protect these vestiges of wild America, Congress placed a number of sensible regulations on wilderness activities. Mechanized travel, for example, was excluded from these areas; those wanting to travel in a wilderness zone had to do so on foot or on horseback.

F. Dale Robertson, chief of the U.S. Forest Service, apparently needs a refresher course on the purpose of the wilderness areas under his jurisdiction. Unilaterally overruling one of his forest supervisors, he has decided to permit the aerial hunting of coyotes in the Mt. Naomi Wilderness Area in the Wasatch Mountains of Utah. Robertson's ruling permits hunters, typically federal agents, to scour the Mt. Naomi Wilderness Area by helicopter and gun down any coyotes they encounter. The hunting occurs during the winter months because the coyotes are easier to track and spot on the snow. The deep snows of the Wasatch Mountains also greatly slow down the coyotes, making their slaughter that much easier.

Robertson took this action under pressure from the livestock industry, which claims that coyotes are cutting into its profits by killing lambs. In so doing, he ignored the recommendations of an environmental analysis prepared by his own Forest Service in response to a challenge by the Utah Wilderness Association. The environmental analysis recommended cuts in predator-control activities throughout the entire Wasatch-Cache National Forest, of which Mt. Naomi is a part, and the total elimination of aerial hunting in the wilderness area. The Forest Service report also recommended that grazing permittees be required to utilize guard dogs and to document coyote kills of livestock before the government turned to aerial hunting. Susan Giannettino, the supervisor for the Wasatch-Cache National Forest, took the recommendations seriously and banned aerial hunting. "I wanted to make management for aerial control consistent throughout the forest and the United States, given that there is no aerial predator control in any national forest wilderness in the lower 48 states" (quoted by *Environmental News Service,* 23 Jan. 1992, 1).

Harold Selman, the grazing permit holder for the Mt. Naomi region, found it easier to petition Robertson than to employ guard dogs or document suspected coyote kills as requested by the Forest Service. Robertson's quick decision to allow the bloodbath to continue suggests that, from his perspective, Selman made the right decision. Dick Carter, coordinator of the Utah Wilderness Association, the group that originally forced the Forest Service to undertake the environmental analysis, was outraged. "Aerial gunning of coyotes in a wilderness area corrupts the intention of protecting wilderness as a place where natural systems operate 'untrammeled by man' and 'affected primarily by the forces of nature.' Not a shred of evidence can be discovered to justify why this wilderness needs aerial gunning, except that the chief was intimidated by the livestock industry" (*Environmental News Service,* 23 Jan. 1992, 1).

For decades coyotes have been under attack by the federal government. Despite their reputation, coyotes are not the "no-account vermin" they have often been portrayed to be. They lead a highly social life, help ranchers by killing rabbits and rodents, can live up to eighteen years, and are capable of running more than forty miles per hour. Estimates are that federal agents throughout the West have killed upwards of five million coyotes since 1919, with untold millions more massacred by private ranchers. Donald Basher, a predator-control expert with the Fish and Wildlife Service, estimates that as many as a half million coyotes are currently killed yearly. The other important biological point that has been lost in Robertson's action is that predators play a very important role in all ecosystems. The balance of nature is shifted dramatically when a natural predator such as the coyote is systematically removed from a habitat, and the results can often be quite destabilizing.

The second situation in which the demands of the grazing industry are creating serious environmental problems brings us full circle to the process of desertification and environmental modification which was discussed at the beginning of this chapter. The operations of the National Park Service and the National Forest Service receive quite a bit of public scrutiny (as they have in this chapter), but the agency responsible for managing more than one-third of all public lands in this country operates with virtually no attention from the public. As a consequence, much of the more than 220 million acres under the control of the Bureau of Land Management (BLM) is in terrible shape, severely overgrazed and overmined.

One of the BLM's biggest faults is that its actions accommodate mining and grazing interests much more readily than they meet the needs of hikers, conservationists, and ecologists. Although vocal charges of this sort come regularly from the environmental community, environmentalists are not alone in recognizing the problem. A report published by the General Accounting Office (GAO), the investigative arm of Congress, levels a striking accusation against the agency. "The BLM has often placed the needs of commercial interests such as livestock permitees [sic] and mine operators ahead of other users. . . . As a result, some permitees have come to view the use of these lands as a property right for private benefit." Given that the GAO can hardly be considered to be part of the environmental lobby, its comments are indicative of a very serious problem.

The situation is not particularly surprising given the origins of the BLM. Formed in 1946, ostensibly to protect open rangeland from overgrazing, the BLM is the offspring of two federal agencies: the Grazing Service, devoted to protecting the interests of the grazing industry, and the General Land Office, an organization charged with the task of disposing of public land. Given such roots, it is almost understandable that, although a 1984 survey done by the BLM itself recognized serious overgrazing on much land under its control, the GAO found that on the vast majority of its land, BLM officials were unable to recommend livestock reductions in the face of political and industry pressure. In addition, the grazing that is permitted is allowed at ridiculously low prices. Fees for grazing BLM land typically run one-quarter of those for similar private land. Grazing fees are so low that they do not even cover the cost of administering the grazing portion of the agency's operation.

With a cattleman in the white house, the Reagan administration was not apt to encourage the BLM to take important environmental steps. Indeed, during the Reagan years, the BLM's planning budget was cut by 50 percent, and although the agency controls approximately 30 percent more land than does the Forest Service, it does so with one-third the staff and half the dollars. Furthermore, the limited dollars that have been requested are very oddly allocated. More than 4.5 times as much money in 1991 was requested for grazing, mining, and timber management than for wildlife management, recreation, and soil and water conservation. Even without additional money, a great deal of environmental good could be accomplished if the budgeted dollars were redirected toward higher-priority items.

Time is quickly running out. Mismanagement is taking an increasingly big toll on fragile ecosystems and their indigenous species. The desert tortoise, for example, a magnificent species found in California deserts, is nearing extinction. Its primary pressure comes from domestic cattle, with whom it competes for food. As overgrazing continues, the tortoise suffers. The numbers speak wonders about the priorities of the BLM: a single tortoise can survive on only twenty-three pounds of plant material per year; a cow and calf together require about twelve thousand pounds. As a federal agency charged with protecting the environment, the BLM should be able to recognize and accommodate this sort of discrepancy. With all that extra grazing and all the associated hooves, arid and semiarid land is being denuded of vegetation, and soil is being compacted. In short, desertification is occurring.

As we overmine, overgraze, and overlumber, and as we construct an unnecessarily complex array of roads through once roadless areas, we fragment ecosystems and weaken the fragile web of species interactions which holds the fabric of natural communities together. Indeed, the single greatest cause of extinctions today is habitat destruction. Although our land use problems are huge, a way of dealing with them was articulately set forth more than forty years ago by the wonderful writer and conservationist Aldo Leopold:

> Quit thinking about decent land-use as solely an economic problem. Examine each question in terms of what is ethically and esthetically right, as well as what is economically expedient. A thing is right when it tends to preserve the integrity, stability, and beauty of the biotic community. It is wrong when it tends otherwise.

These are words that modern-day environmentalists would do well to heed. If more people learned to ask questions about the effects of various actions on the intrinsic health of ecosystems, and if they learned to incorporate answers to those questions into the formation of land use policy, our land and, therefore, we and our children would be much better off.

Chapter Ten

༓

Myths
of the
Technological
Fix

Partly as a result of the terror fueled by the potency of the super-powers' nuclear arsenals, partly because of the fear generated by nuclear accidents like those at Three Mile Island and Chernobyl, partly because of the growing concern in the minds of many that scientific fraud is rampant, and partly because of the humiliation heaped on scientists by Senator William Proxmire and his Golden Fleece award, we Americans no longer hold the scientific establishment in the same esteem we once did. Even as our faith slips, however, there remains the strong belief that science and technology will rescue us from toxins. Rescue us from illness. Rescue us from global destruction. Too many of us too blithely assume that we need not deal with the base causes of our environmental problems because soon-to-be-discovered technological solutions will make those problems obsolete. This unhealthy and unrealistic viewpoint all too often diverts our attention from meaningful problem solving and instead centers it on all too facile solutions.

On one level, the science-to-the-rescue attitude focuses far too much attention on applied rather than on pure research. Enjoying such a mind-set, our elected officials think the type of scientific investigation which is worthy of funding is that which is designed to solve problems. Pure research, research that simply advances a scientific discipline for its own sake rather than providing a solution to a specific problem, is often viewed as a luxury. But as we saw in Chapter 4, without a foundation of pure research, the work of the applied scientists would quickly grind to a halt. And what is most important to recognize is that the applied spinoffs from pure research are almost always impossible to predict in advance.

On another level, the myth that there will always be a technological solution to our environmental problems focuses our attention on complex rather than simple solutions. Barry Commoner, biologist, environmental activist, and onetime presidential candidate, has spent a large portion of his illustrious career pointing out the difference between pollution prevention and pollution control. His message is so simple that it has been largely ignored. We would be wise to review his ideas, set forth most recently in his 1990 book entitled *Making Peace with the Planet:*

> A control device always allows some pollution to enter the environment, so that increased productive activity negates the device's intended effect. In contrast, when a pollutant is simply eliminated or banned, its rate of entry into the environment falls permanently to zero. If pollution is prevented, environmental quality is compatible with increased economic activity; if pollution is controlled, they clash. In the task of restoring environmental quality, prevention works; control does not. . . . (44) Only in the few instances in which the technology of production has been changed—by eliminating lead from gasoline, mercury from chlorine production, DDT from agriculture, PCB from the electrical industry, and atmospheric nuclear explosions from the military enterprise—has the environment been substantially improved. When a pollutant is attacked at the point of origin—in the production process that generates it—the pollutant can be eliminated; once it is produced, it is too late. This is the simple but powerful lesson of the two decades of intense but largely futile effort to improve the quality of the environment. (55)

Even the successful cases of pollution prevention cited by Dr. Commoner take longer to be fully effective because of the staying power of many of the chemicals in question. As was pointed out in Chapters 5 and 8, some chemicals banned decades ago are still posing health problems.

A related reason why pollution control isn't nearly as effective as is pollution prevention stems from the population growth patterns discussed in Chapter 5. As the planet's human population continues to double at a seemingly inexorable rate, it becomes increasingly difficult to control the total amount of pollution produced. The automobile in America clearly demonstrates the problem. Although the United States' automobile population has been increasing significantly faster than has its human population, let us assume an equal growth rate for the two. Let us assume further that a major breakthrough in technology yields a reduction in emissions from each new car by 15 percent. How much cleaner will American skies become because of this technological advance? The disappointing answer is that we won't be better off than we currently are. For one thing, it will take quite a few years for the country's automobile fleet to be replaced by the new models. Furthermore, as the number of automobiles increases, the total amount of pollution produced will rise even as the amount per car decreases. Do not misunderstand me; a 15 percent reduction in automotive exhaust would be most welcome—it would surely slow the rate at which our skies become ever more polluted—but such pollution control approaches can never be very effective in the face of a growing population.

As Commoner says, we must think instead of pollution prevention. A high-level panel of scientific experts, ignoring his simple advice, has come up with what it thinks might be a dramatic cure for global warming. However, although the panel's solution is in keeping with a basic law of ecology, its idea runs afoul of an equally basic law of medicine. The experts, a panel of top scientists gathered by the National Research Council (NRC), the investigative arm of the U.S. National Academy of Sciences, the National Academy of Engineering, and the Institute of Medicine, have endorsed a plan to fertilize the planet's oceans with iron. Because iron is supposed to promote the growth of tiny marine algae known as phytoplankton, and because phytoplankton, like all green plants, take up carbon dioxide, one of the most troublesome greenhouse gases, and release oxygen, it is thought that a dramatic increase in phytoplankton will lead to the removal from the atmosphere of a large percentage of the offending carbon dioxide.

Why iron? As every gardener knows, and as I have taught my introductory ecology students early in every semester, the answer is quite simple. In 1840 the great German chemist Justice von Liebig conducted a series of experiments and published a scientific paper that established the "Law of the Minimum." That ecological axiom states that the "growth of a plant depends on the amount of food-stuff which is presented to it in limited quantity." Even more simply, plant growth is always limited by a particular environmental factor. Add more of that factor, and plants will grow until some other factor becomes limiting. The NRC scientists, recognizing that iron appears to be the factor limiting the growth of phytoplankton, want to add more iron to the phytoplankton's habitat.

So confident are the scientists of their solution that one of the leading proponents of this remedy, John Martin of Moss Landing Marine Laboratories in California, has been quoted in the *Washington Post* as saying, "You give me half a tanker full of iron, I'll give you another ice age." Needless to say, no one on the NRC panel wants quite that much iron, and, also needless to say, no purposeful human manipulation of the natural environment of this magnitude has ever before been undertaken. Because the phytoplankton are at the bottom of the ocean's food chain, fed on by tiny zooplankton, which in turn are eaten by the fish and mammals of the sea, the ecological consequences of the addition of iron might be enormous. This uncertainty alone might be reason enough not to pursue such a massive manipulation.

There is a far better reason, however, not to proceed along the lines of the NRC panel's recommendations, and that reason comes from medicine rather than ecology. Doctors have long recognized that it is far less productive to treat the symptoms of a disease rather than its underlying causes. Aspirin, for example, might be wonderful at bringing down a fever caused by a bacterial infection, but it will not help remove the offending bacteria. With the fever reduced, the patient may seem to be improving only to suffer a massive and perhaps deadly relapse when the bacteria reach immense, uncontrollable quantities. The fever itself can help to kill the bacteria, and it can serve as a warning device signaling significant changes in the health of the patient. By treating the symptom, all of its beneficial effects are lost.

Fertilizing the oceans with iron is like feeding a sick patient aspirin. Increased quantities of phytoplankton might well reduce the amount of carbon dioxide in the atmosphere, but they will not get at the root cause of global warming. Moreover, given the politics, both national and international, of attempting to rein in emissions of offending gases, any program that appears to make such control less immediately pressing is sure to add ammunition to those who would rather not take any action at all. As with an untreated bacterial infection, continuing to use our atmosphere as a repository of unwanted pollutants is a prescription for suicide, particularly so in light of recent research indicating that a significant bout of global warming might be triggered more suddenly than previously thought.

Regardless of how much of an increase in phytoplankton growth might be coaxed out of our oceans, the amount of carbon dioxide which will be absorbed will never be infinite. Ultimately, then, we will still need to control our profligate habits if we expect to live in harmony on this planet. Fertilizing the oceans, while delaying the problem, will increase its magnitude and ultimately will make finding an acceptable solution even more difficult than it is already. By delaying, we are doing what we do so well, bequeathing to our children our own most difficult problems.

We also need to be especially vigilant to be sure that when a technological solution for an environmental problem is proposed, it doesn't, in turn, create an even bigger problem. We stand on the brink of repeating what Ellen Silbergeld, toxicologist, chief scientist for the Environmental Defense Fund, professor at the University of Maryland, and MacArthur Fellowship–winner, has called the "major environmental disaster of our time." We might well step over that edge with industry's assurance that environmentally we are doing the correct thing.

More than five years ago, Silbergeld spoke those words in reference to lead pollution caused predominantly by leaded gasoline. She went on to say, "Lead poisoning is one of the five most common diseases of children. It is entirely preventable: we possess the knowledge and resources to prevent it." Despite this call to action, and despite EPA statistics demonstrating that at least 7,000 cases of lead poisoning in children, 387 deaths, and 123,000 cases of high blood pressure could be avoided annually by enforcing a proposed ban on leaded gasoline, the federal government under the Reagan administration opted to delay such a ban.

This history is of critical importance because the people who brought us leaded gasoline, the Ethyl Corporation of America, have new plans. They want to market a new gasoline additive, HiTec 3000. Advertising executives for the Ethyl Corporation are attempting to seize the environmental high ground for their new product by claiming that by reducing emissions of hydrocarbons and nitrous oxides from cars, HiTec 3000 will yield significant environmental benefits. What they fail to highlight, however, is that HiTec 3000 contains the heavy metal manganese and that its widespread use will release large quantities of this toxic element into the atmosphere.

The lack of any environmental awareness by Ethyl Corporation in these times of heightened public sensitivity is amazing. Apparently the captains of this particular industry imagine that today they can get away with the same sort of shoddy research they conducted in 1925. It was seventy years ago, after all, that the Ethyl Corporation argued that the amounts of lead they planned to add to gasoline were negligible and that, in any case, lead was only toxic at the high doses encountered in certain industrial settings. Today's proposal for HiTec 3000 makes exactly the same argument for manganese—manganese emissions will be negligible, and the metal is only toxic at the high doses encountered in certain industrial settings. We now know just how foolish the 1925 claims were. Leaded gasoline created a health hazard of shocking proportions from which we are still recovering. Significant sums of money are still needed to clean up the residues of lead fallout on playgrounds and roadsides.

There is ample evidence suggesting that, despite the media campaign waged by the Ethyl Corporation, the situation will be no better with respect to manganese. Experts agree that at high doses manganese is a human neurotoxin causing persistent and irreversible pathological effects on brain structure leading to severe impairments in movement and mental state. There are also indications that high levels of manganese can have deleterious effects on the development of fetuses and on young, growing children. Although it is unclear whether or not manganese is carcinogenic, studies have strongly suggested that it can cause breakage of DNA, which may be indicative of cancer-causing potential. Unfortunately, no studies have been performed which allow predictions to be made about the long-term, chronic, low-dosage health effects of manganese. Being so foolish as to permit HiTec 3000's use as a gasoline additive will demonstrate what the long-term consequences of manganese exposure are in exactly the same way that leaded gas so dramatically forced us to recognize the terrible consequences of lead. This type of large-scale experiment on the human race should not be permitted to recur. We should not even be fooled into thinking that we can try HiTec 3000 for a limited time. Again, remember our experience with lead. Even after the need for lead in gasoline was being phased out by automobile manufacturers, and even after its deadly legacy had been fully appreciated, the federal government proved unable to stand up to the gasoline industry and ban the product domestically.

It is worth taking a moment to explore the situation of leaded gasoline in the developing world. As Barry Commoner claims, the phasing out of leaded gasoline has been a huge environmental and public health success. With the demise of leaded gas, we have averted hundreds of thousands of cases of lead poisoning in children and thousands of deaths. Lead is, after all, responsible for causing brain damage, neurological dysfunction, kidney damage, and exacerbated cases of high blood pressure. Lead poisoning also dramatically decreases intelligence levels of children (and consequently of adults). Those figures and those effects are the consensus of doctors, environmental groups, and the Environmental Protection Agency. The patterns in New York City alone tell a dramatic story. Sergio Piomelli, a hematologist at Columbia University's Children's Hospital, said that before the phaseout of leaded gas 30 percent of the city's children showed elevated blood lead levels; afterward only 1.5 percent showed levels that high. That, at least, is the story in the industrial world. In developing countries the situation is strikingly different.

Mario Epelman, a physician with Greenpeace's Latin America Project, put it best when he said, "Today, we have one gasoline for the rich countries, and another, deadlier gasoline for less industrialized countries" (*Greenpeace*, Oct.–Dec. 1991, 18). Leaded gas, cheaper than unleaded, is still widely used in the developing world, where, consequently, ambient levels of lead greatly exceed what the industrialized countries have decided is safe for their citizens. The situation is analogous to that of the pesticides banned throughout much of the industrialized world. Manufacturers, rather than completely halting production of toxic but profitable products, simply look to sell those products to overseas markets, where environmental and health restrictions are much reduced. The Ethyl Corporation, based in Richmond, Virginia, but producing its deadly product in Ontario, Canada, supplies a good portion of the Third World with tetraethyl lead (TEL) and its attendant health problems.

Although TEL has been banned for domestic use in Canada, it is permitted to be manufactured for sale abroad. The attitude of Ethyl officials to this situation is appalling. John Street, Ethyl's Canadian environmental and safety manager, suffers no moral qualms about being associated with the production and exportation of a product that has been banned in part of the industrialized world. He rationalizes that if TEL were not made in Canada, it would be produced elsewhere. Floyd Gottwald, CEO of Ethyl, has actually gone on the offensive, claiming in a July 1991 letter that "no conclusive scientific evidence has ever linked the use of small amounts of TEL in gasoline to human health problems. This product is safe when used properly." Given the number of studies showing exactly the opposite and the broad consensus among scientists and medical doctors about the dangers of TEL, Gottwald sounds about as believable as tobacco industry spokespeople claiming that cigarette smoking has not been proven dangerous.

Even Gilbert Grosvenor, an Ethyl board member and president of the National Geographic Society, which has for years attempted to awaken our understanding of Third World predicaments, has attempted to defend the use of TEL, callously claiming in a letter in July 1991 that Ethyl is actually helping the world's poorest countries because "elimination of this additive could cause significant damage to automobiles and thus to the transportation systems and economies of those countries." Although it has long been believed that unleaded gas might damage valves in older cars, tests by both the U.S. Army and the U.S. Postal Service have found that with normal usage no such damage occurs. All available evidence instead points to a single conclusion: Ethyl Corporation officials remain arrogant, heartless, and self-serving in their defense of their product. With such an attitude, how wise are we to believe their claims that HiTec 3000 will be an environmentally clean and healthy solution to the part of global warming attributable to automobile exhaust?

We especially need to be wary of high-tech environmental solutions that appear to be a panacea for serious problems. One such false panacea is biodegradable plastic. A growing industry, spurred on by an aggressive corn lobby anxious to find expanded markets for its products, is pressing the environmental advantages of using plastic products that are actually a combination of plastic and corn starch. The argument seems straightforward enough: because biodegradable plastics decompose fairly readily, increasing their use will, among other things, help us reduce litter and extend the life of our landfills. Happily, not everyone is buying the industry's rhetoric. Richard A. Denison, a biochemist working for the Environmental Defense Fund, goes so far as to say that biodegradable plastic may actually be a greater environmental threat than is regular plastic. His compelling and somewhat surprising argument comes in four parts.

First, Denison claims that it is highly unlikely that most biodegradable plastic will ever actually biodegrade when it is disposed of properly. Because biodegradable plastics come in either one of two forms, photodegradable or biodegradable, they require either light or oxygen, respectively, for the desirable breakdown to occur. When such products are buried in landfills, however, they are exposed to neither and will not, therefore, degrade. Denison points out that excavations have found that even such fragile organic items as hot dogs, banana peels, and newspapers are recognizable decades after burial.

Oddly enough, the fact that the large majority of biodegradable plastics will never degrade is, in Denison's mind, the good news. Even if the plastic *did* degrade, our problems would just be beginning. Denison points out that the plastic portion of the product does not actually change when degradation occurs. Rather, the cornstarch or other vegetable matter used as a matrix is what breaks down, leaving behind a fine powder of "plastic dust." It is this dust that Denison fears. Because numerous toxic additives such as lead and cadmium are routinely used in the manufacture of plastic, Denison worries that the plastic dust will be laced with these toxins. The poisonous chemicals are far more mobile and are thus more likely to enter our food chain and water supply when they are in small dust-sized particles rather than relatively inert pieces of packaging material.

Denison's third concern is economic. Increasingly, businesses are finding a market for used plastic. Through recycling, old plastics can efficiently be converted into a wide range of durable items such as plastic lumber, piping, carpeting, and textiles. Indeed, many environmentalists see such plastic recycling as one of the most promising solutions to our growing waste problem. An expanded biodegradable plastic market might well severely wound the plastics recycling industry by compromising the structural integrity of the items produced. Obviously any structure constructed with materials liberally laced with biodegradable plastic cannot be expected to be usable for very long. Additionally, it seems reasonable to suppose that consumers will be reluctant to participate in any plastic recycling program if forced to sort biodegradable plastics from normal ones.

Finally, Denison is concerned that a turn toward biodegradable plastic will lead to the consumption of additional natural resources. He argues, for example, that significantly more plastic resin is required to make biodegradable items than is required to make ordinary plastic items of equal strength. Perhaps even more important, Denison believes that the presence of biodegradable plastic will undermine whatever growing conservation ethic might be spreading among consumers. Because users will be led to believe that biodegradable products will simply break down and return to nature, they will be less likely to concern themselves with the amount of plastic they throw away.

There is a world of difference between recyclable and biodegradable plastic. We should not allow possible consumer confusion between the two to undermine the important efforts being made by the recycling industry. Similarly, we should not allow the fact that biodegradable products sound so environmentally respectable to lure us into accepting their widespread use without critical inquiry. It is unfortunate, but nevertheless true, that biodegradable plastic is not the environmental panacea that some proponents would like us to believe.

In an attempt to find a market niche, companies occasionally market high-tech solutions to environmental problems that don't even exist. By craftily preying on our fears, manufacturers often turn a handsome profit. One of our most common fears centers on insects. When it comes to these six-legged organisms, many people become downright irrational in their distaste and loathing. Many insects do spread disease and discomfort as well as destroy crops, but many more provide humans with a wealth of services. From preying on pests to pollinating crops, millions of insects provide benefits that we would be hard pressed to do without. Nonetheless, people have spent untold billions of dollars in an attempt to eradicate insects.

This massive disgust has led to two problems. First, most extermination efforts are indiscriminate. Rarely are specific species targeted—typically a monarch butterfly is equally at risk as a mosquito. Second, unscrupulous entrepreneurs have homed in on our hatred and have marketed products making wild claims with no scientific support. It was not very long ago that large sums of money were wasted on ultrasonic devices supposedly designed to drive insects and rodents from consumers' homes. When test after test failed to demonstrate any decrease in pest populations, the Environmental Protection Agency and the Federal Trade Commission began to take legal action against the manufacturers of ultrasonic devices.

Neither agency, however, has yet seen fit to take action against the electronic bug zapper, another product whose efficacy is suspect but whose sales have been dramatically increasing. Homeowners are even more easily taken in by the bug zapper than they were by the ultrasonic devices because the carcasses of incinerated insects offer tangible proof that insect populations are declining. Auditory proof, in the form of the incessant "zzzzap," is also plentiful. Given all of this evidence, there can be no controversy over the fact that bug zappers kill insects. The complication is that the zappers are not killing the insects that the millions of purchasers think they are. The zappers are primarily marketed as an environmentally sound way to reduce the populations of biting insects that pester humans. But by luring insects with an ultraviolet light, the zappers work only between dusk and dawn, a time when most bloodsucking insects are least active. Furthermore, biting insects, like mosquitoes, rarely orient to light and only to a limited extent use visual cues to find hosts. Instead, they are drawn by smell to the carbon dioxide exhaled and to the chemicals in mammalian sweat. The only mosquitoes or biting flies likely to be zapped by such a device are those who unwittingly encounter the machine while searching for a bite to eat. Not surprisingly, the controlled studies that have been performed have found that very few biting insects are being killed by zappers and that the incidence of mosquito bites in "protected" yards equals that in "unprotected" yards.

Bilking gullible homeowners with false advertising should not be permitted, but the more important reason to educate the public about the reality of bug zappers is the millions of harmless and, indeed, helpful insect victims incinerated. The organisms most likely to be singed are those nocturnal and crepuscular (active during dusk and dawn) ones that orient to light and are thus drawn to the ultraviolet rays. Many of these are nectar-feeding moths that are essential for the pollination of a wide array of large, showy flowers. A zapper in a garden one summer might well reduce the number of flowers blooming the next. With sales in the millions of units, we might also soon discern a decrease in the populations of insectivorous birds who prefer their prey uncooked.

Zappers have one other unintended effect. As the insects in a yard with a zapper are killed, many others move in to take their place. In short, although zappers are demonstrably reducing insect populations, they are not necessarily doing so on a scale observable by the homeowner who has purchased the zapper and who is footing the electric bill.

The biology of biting insects makes it clear that zappers cannot be a particularly effective means of control, yet an ignorant public is increasingly buying the zapper scam. Ironically, the ultrasonic pest control devices that most people now recognize as worthless are looking better and better in comparison. Since they do not kill anything, they do not have the potential to cause as much environmental chaos as do the zappers. And at least they are silent and do not disturb the neighbors.

From zapping insects in our backyards to adding iron to the planet's oceans, we don't seem to give much thought to the long-term consequences of altering entire ecosystems when in search of a technological solution to a problem. Instead, we focus on the immediate problem and forget another basic law of ecology: all living things are interconnected. Sometimes those links are very tight. As discussed in Chapter 6, when the huge, flightless dodo was driven to extinction in 1675, the tambalocoque, *Calvaria major,* a large tree species endemic to the island of Mauritius, began a seemingly inexorable decline into oblivion because its seeds depended on passage through the gut of the dodo before germination could occur. Sometimes, as with the destruction of Caribbean coral reefs caused by deforestation of the Amazon basin, discussed in Chapter 5, those links are more diffuse but nonetheless very real.

Consider a plan promoted during the Reagan administration's war on drugs. As we have so tragically witnessed in Vietnam, Central America, the Persian Gulf, and Bosnia, war wreaks havoc on the environment. Had the Reagan administration had its way, its war on drugs would have had a similar effect. The plan was elegant indeed. Officials proposed to use an herbicide to stop the growth of coca plants in Peru's main cocaine-producing regions. There were two main problems with the plan, one practical and one of a biological nature.

From a practical perspective the plan was in trouble because Eli Lilly and Company was uninterested in selling the government large stocks of its herbicide, tebuthiuron, marketed domestically under the brand name Spike. Although the company was unwilling to discuss the issue in any detail, two concerns appear to have been primary. The herbicide, approved for use in the United States to clear brushland and on noncrop pastureland, has not been widely tested under the tropical environmental conditions widespread in Peru. The company, in a situation reminiscent of a test program in Colombia in 1984, might well have been worried about the legal problems associated with the use of such a toxic product under untried conditions. At that time, Colombian lowlands were sprayed with Garlon-4 to see if the herbicide might eradicate the highly resistant coca plants, but the project was abandoned after the Dow Chemical Company, the maker of the herbicide, demanded indemnification against lawsuits arising from the use of the chemical. (It is interesting that in this case herbicide manufacturers were concerned about the environmental conditions under which their products are used, but, as we saw in the discussion of Susan Fisher's work in Chapter 7, they, along with the EPA, seem to dismiss similar domestic worries.) In addition to the potential for massive litigation, Lilly might also have been concerned about possible violent retaliation by members of the international drug cartels should Spike prove effective in eradicating coca plants.

From the biological perspective, the plan was troublesome because Spike is a nonspecific herbicide that kills woody and herbaceous vegetation alike, with a single application remaining effective for up to three years. In the United States the chemical is not approved for use on cropland because such use would keep all food plants from growing for an extended period. In Peru, where coca plants are typically interspersed among both agricultural fields and native vegetation, aerial spraying of Spike would necessarily contaminate more than just target plants. Local farmers might thus be unfairly asked to bear the burden of our country's drug eradication program. Native vegetation, also at risk from the proposed spraying program, is composed of unique remnants of tropical rain forest. The Upper Huallaga Valley amid the Andes Mountains is, in addition to one of the world's coca centers, a botanical and zoological treasure. Because the valley is isolated by high mountains, it contains many rare and endangered plant and animal species, most of which have not yet been cataloged.

Equally troublesome was that the State Department was pushing the spraying program having performed very little environmental testing. Although State Department official Ann B. Wrobleski was quoted in *Science* as saying that Spike "is less toxic than aspirin, nicotine and nitrate fertilizers," the EPA stated quite clearly that it would not vouch for the safety of a product used in a foreign country or in an environment other than that for which the herbicide was originally registered.

Similarly, the U.S. Department of Agriculture did not want to get involved, saying that environmental assessments of herbicides are not included in its responsibilities. Even so, the State Department was actually suggesting that Spike might be a tool for improving the environment. The logic was that since coca growers cause severe environmental degradation, anything that stops them will be environmentally beneficial. Half of that logic is impeccable—coca growers have deforested large tracts of native forest, where erosion quickly becomes a serious problem. Additionally, the growers have been regularly reported to pollute streams with the kerosene, sulfuric acid, acetone, and toluene used in the processing of coca leaves. However, the remaining half of the State Department logic leaves a bit to be desired. Although an effective coca eradication program will probably put an end to much of the environmental degradation caused by the drug makers, the program as designed would probably have thoroughly destroyed the environment in the process.

The coca eradication plan arose from the Reagan administration's view that it would be easier to control the production of cocaine abroad than to design creative ways to limit domestic demand for the drug. The view is ironic in light of data released by the Drug Enforcement Administration at the time indicating that the entire U.S. demand for cocaine could easily be provided by an area less than ten miles square. In any case, only the most imperialistic government would risk destruction of a foreign country's fragile tropical habitat rather than deal with the root of the problem at home.

But then Ronald Reagan never evidenced any great fondness for trees. From his famous statement about redwoods when he was governor of California ("If you've seen one redwood . . .") to his pronouncement on the presidential campaign trail that trees are notorious polluters, he set a tone for the destruction of forests worldwide. We've come a long way from those days—maybe too far. Almost everyone in sight these days is promoting the planting of trees as a means of slowing the rising tide of greenhouse gases. The problem is that almost everyone is doing it in what must be considered an environmentally blind fashion.

In a stinging indictment of the tree-planting movement, controversial and articulate *Audubon* magazine columnist Ted Williams claims that tree promoters are actually doing more harm than good. Speaking of average citizens who have been swept along by the tree-planting bandwagon, Williams asserts, "It's not that tree-planters don't do lots of good by frequently planting trees in the right places, and it's not that they aren't nice people who mean well. It's just that, in their innocence, they are an environmental menace." Their confusion falls into three major categories: disrupting natural ecosystems by planting nonnative species; destroying natural ecosystems by planting trees where, ecologically, none were meant to be; and providing the public with terribly erroneous information in the guise of ecological advice. All three problems are more significant than they might appear at first glance.

While it is surely true that trees are capable of absorbing quite a bit of the most abundant greenhouse gas, carbon dioxide, the wrong tree in the wrong place is capable of causing untold amounts of environmental damage. For the past seventy or so years, ecologists have known that when nonnative species are introduced into new environments, typically without the predators, parasites, and competitors with which they coevolved for thousands of generations, they often outcompete many of the native species and become pests. The blue gum eucalyptus that was introduced into California from Tasmania has spread so well that it has overrun entire native ecosystems, endangering many plants and animals. Huge numbers of the trees that are being pushed are alien to North America, perhaps the most common of which is the Afghan pine. A number of thoughtful local environmental organizations have agreed to participate with the national tree-planting groups with the guarantee that only exclusively indigenous species be promoted. No such assurances have been forthcoming.

Even when native species are planted, some habitats are ecologically inappropriate for their introduction. The tall-grass prairie, stretching from Ohio to western Kansas, although naturally devoid of trees, is a magnificent ecosystem. Nonetheless, Julius Sterling Morton, the founder of the National Arbor Day Foundation, started the group in 1870 to create "a grand army of husbandmen . . . to battle against the timberless prairies." The organization, as if in bizarre misappropriation of Freud's famed penis-envy theory, still espouses this strange goal. Similarly, many naturally treeless hillsides in San Francisco have been blanketed with nonnative trees while the original native grassland-wildflower communities have been forever lost.

The tree planters have also set their sights on environments that typically have trees but currently do not. The problem is that too frequently these treeless areas achieved their present, naked state as a result of natural environmental occurrences, such as fires and floods. The normal sequence of events is for these naturally disturbed environments to undergo a successionary process whereby a full complement of plants will eventually be reestablished. Rushing in with trees, either native or alien, only disrupts the natural process and makes it less likely that a stable community will mature. Yet the American Forestry Association–sponsored Global ReLeaf organization has done just this by planting 137,500 trees in South Carolina's Francis Marion National Forest in the wake of Hurricane Hugo. Additionally, despite wise resistance from the National Park Service, the National Arbor Day Foundation has been advertising that "millions of trees must be planted" to help Yellowstone recover from its 1988 fires. The misleading message from these misguided efforts is that natural processes are incapable of creating a stable ecological balance and that only human efforts are effective at doing so.

Planting trees can be an environmentally sound thing to do. But the specific trees and the specific locations must be chosen with care, knowledge, and respect. Environmental interactions are beautifully complex, and heedless tree planting too often leads to environmental havoc.

One final, chilling demonstration of what might happen when we fail to analyze fully the consequences of our actions can be found in a report released a couple of years ago by the U.S. Congress's Office of Technology Assessment (OTA). The report, entitled "The Electronic Supervisor," focused primarily on technology used to monitor the performance of workers. Much of the report discussed the uses and implications of technologies already in place, technology that permits the monitoring of telephone conversations between workers and clients by supervisors, as well as technology which, via computer, records the number of keystrokes per minute typed by workers using word processors. But the government's report surpassed this by now mundane, if still insidious, technology and introduced a number of additional methods for monitoring worker performance. Among these are polygraphs ("lie detectors"), substance abuse tests, and genetic screening.

The first two are currently in widespread use even though there has been enormous disagreement concerning their liability, as was discussed in Chapters 2 and 3, respectively. Additionally, both have generated debate with respect to the legal implications associated with the necessary compromise of workers' Fifth Amendment rights and the stress induced in workers subjected to such testing. Nonetheless, it is genetic screening that should worry us most.

According to the OTA, a small number of companies currently use genetic screening techniques in hiring, and a growing number seem interested in doing so. One goal of this sort of screening is to allow the employer to differentiate and eliminate those job candidates perceived to be either unhealthy or potentially unhealthy. When the company contributes to health and life insurance payments, the employer might enjoy lower group rates and save substantial amounts of money by hiring only workers without the genetic predisposition to certain diseases. As genetic tests become more sophisticated, employers also hope to be able to determine which potential employees are most likely to suffer negative reactions to environmental hazards on the job. If, for example, a test was developed to indicate susceptibility to chlordane, pesticide manufacturers could employ only the least susceptible individuals.

Both of these goals have already drawn significant criticism. Instead of hiring workers least sensitive to toxins, opponents argue that companies should redesign manufacturing processes to minimize the production of toxins and fashion safe disposal procedures. In short, genetic screening associated with selective hiring may provide employers with an opportunity to avoid the responsibility of designing safe workplaces, to the detriment of society in general. In addition, selective hiring may create different classes of workers. Employers offering jobs requiring extensive training, both in terms of money and time, will want to maximize their profits by hiring only workers likely to remain healthy for extended periods. A company might refuse to hire individuals showing the genetic predisposition for certain disabling diseases, for example. Employers offering more menial jobs requiring a minimal amount of training, on the other hand, may prefer workers whose genetic profiles suggest short life expectancies. Under such conditions companies might minimize the money they pay in pensions. Although the technology does not yet exist to perform most of the genetic tests employers might want, advances in recombinant DNA technology along with the massive effort being expended in the Human Genome project to identify all human genes suggest that such tests await us on the not-too-distant horizon.

The ethical and legal implications of genetic screening are staggering. As Dorothy Nelkin and Laurence Tancredi wrote in their 1989 book entitled *Dangerous Diagnostics: The Social Power of Biological In-*

formation, if we do not face these issues directly, we "risk increasing the number of people defined as unemployable, uneducable, or uninsurable. We risk creating a biologic underclass." Now is the time, before the bulk of the tests are commercially available and momentum makes discussion much more difficult, to discuss their use in the workplace. Now is the time to formulate regulations to protect ourselves from possible abuses of this technology. For only after open, informed discussion has occurred and sound laws are in place can we be assured that future advances in technology will not bring about a retreat in civil liberties.

That is a lesson that we would be wise to extrapolate to so many other public policy decisions. It is far easier never to start certain activities than to stop them. Although it is easy to be lured into a false sense of security by some of the easy answers provided by technology, we will not necessarily be better off if we succumb to its allure. The situation is analogous to so many of those quick weight loss schemes that promise to melt pounds away without any effort. Although a short-term advantage might be gained under such a regimen, the pounds are sure to return in the absence of increased exercise and a modified diet—along with a loss of self-esteem for having failed to keep the weight off. So it is with seemingly easy technological fixes to some of our global problems. We might genetically screen workers rather than clean up the workplace, seed the ocean with iron rather than reduce our emissions of greenhouse gases, or turn to biodegradable plastic rather than create an environmental ethic that considers littering abhorrent and recycling standard, but by doing so we would be hiding from our real problems. Meaningful cures will come only when we squarely face the underlying causes of our problems. That means that we have to begin to address the world's population problem as well as our profligate use of energy and our indiscriminate use of dangerous, and often unnecessary, chemicals. As we will see in the next chapter, technology surely has a role to play in creating such a world, but we need to stop looking toward it as a panacea.

Chapter Eleven

ॐ

Prospects

Medical doctors and patients alike lament the side effects of drugs. More often than not, such complaints are just a matter of perspective. Although the side effects are very real, and often very troubling, they almost invariably come about because the drug in question is doing exactly what it was designed to do; the doctor and the patient, however, are only interested in a subset of its activity. For example, one common, unpleasant result of the administration of broad-spectrum oral antibiotics is diarrhea coupled with intestinal discomfort and reduced nutritional intake. The antibiotics are given to kill a bacterial infection, and usually, leaving aside the growing problem of drug resistance because of overuse of such medicines, the antibiotics do exactly that—kill bacteria. Because so many of our antibiotics are not targeted for specific bacteria, oral doses usually wipe out a significant portion of the intestinal fauna so important in aiding digestion. "Good" and "bad" bacteria alike are killed. In general, the benefits (killing the disease-causing bacteria) outweigh the costs (killing the intestinal bacteria), and thus oral antibiotics remain a common course of treatment.

Competent physicians should be careful to consider all the costs as well as all the benefits before prescribing any course of treatment. With such a mind-set, side effects, rather than seeming to be extraneous factors that appear troublesome after the fact, may be taken into consideration right from the beginning. While this holistic view of medicine will not eliminate the unwanted effects of drugs, both patient and doctor will be sure to encounter fewer unpleasant surprises. Additionally, there will be times when it is clear that the consequences of the negative side effects outweigh the positive, suggesting a modified course of treatment.

Some of our most serious environmental problems arise because economists, politicians, and business leaders, as well as the bulk of the rest of us, have an unrealistic view of side effects. Far too frequently when we make a social decision, we focus our attention too narrowly on the perceived benefits of our actions while neglecting to examine possible negative consequences. As the negatives become farther removed from the positives, they become easier to ignore. But just because those negatives are so easy to ignore in the beginning does not mean that they will be unimportant toward the end. In fact, we pay dearly for our ignorance. Remember that the 1925 decision, as discussed in Chapter 10, to add lead to gasoline to improve performance ignored the health consequences resulting from massive amounts of airborne lead in inner cities. Because of that decision we have experienced uncountable cases of lead poisoning leading to decreased intelligence, high blood pressure, kidney damage, neurological dysfunction, and death.

Economists don't like to talk about side effects; instead, they speak of externalities. When calculating the cost of goods or services, all the variables ignored by economists are called externalities. All the indirect costs (or negative consequences) are considered externalities because they have remained external to the fixing of prices. Like medical side effects, when externalities are taken fully into consideration, better-quality decisions will be made. Economist Alan Krupnick of Resources for the Future has phrased the situation very nicely: "The idea is that if you can add in all of these hidden costs, you give consumers a better idea of the consequences of their actions" (*Science*, 25 June 1993, 1885).

Internalizing externalities is a remarkably conservative thing to do because it relies directly on one of the basic notions of capitalism—that markets respond to price signals. Simply put, if resources are priced to reflect their real costs, a well-functioning capitalistic economy will optimize use of those resources. On the other hand, when the true costs of resources are hidden, ignored completely, or subsidized by the government, we create a situation that encourages inefficient use of resources, leading to exploitation and rampant pollution. Unfortunately for us all, the government is usually remarkably good at ignoring externalities. The present state of public grazing lands discussed in Chapter 9 is a perfect example. Currently, ranchers pay approximately $1.86 per month to graze a cow and a calf (or five sheep) on federal land; similar grazing charges on adjacent private land run up to $15.00 per month. Not surprisingly, federal grasslands are typically overgrazed and experience extensive erosion problems while underutilized private lands enjoy a much healthier environmental state.

A number of studies have shown that we have an enormous distance to go before we are even close to pricing resources meaningfully. An analysis performed by three economists, James Costanza from the University of Maryland, Stephen Farber from Louisiana State University, and Judith Maxwell from Ohio State University, is an excellent case in point. Currently, ignoring all externalities, wetlands on the coast of Louisiana are priced between $200 and $400 per acre. Costanza summarizes the results of his group's study by saying, "If you consider the recreational value, storm protection, fisheries output, and trapping, you get a real cost of between $2,400 and $17,000 per acre" (*Science*, 25 June 1993, 1885). Even the Army Corps of Engineers has arrived at a similar conclusion. The corps has estimated that approximately $17 million per year in flood damage is averted by keeping a wetland on the periphery of Boston Harbor free from development.

Economist Harold M. Hubbard from Resources of the Future has taken on the huge task of adding in some of the externalities of the cost of oil. He has estimated that the annual cost of defending the Persian Gulf shipping lanes totals $15 billion, or $23.50 per barrel of imported oil. When he adds in other externalities, such as the detrimental health effects attributable to oil, the hidden costs skyrocket to between $100 billion and $300 billion. Certainly all of us would think much more carefully about our use of oil and might arrive at some surprising decisions if, individually, we were expected to pay the real cost of oil. The point to remember is that the externalities are present and concrete even though we fail to include them in our personal decision making. There can be no argument that collectively we pay a huge yearly defense bill and a huge health bill even as our newly acquired oil forces us to face the grave consequences of global warming. Spreading those costs and consequences out broadly over society does not make them vanish.

Rather, as Garret Hardin so succinctly pointed out years ago in his 1968 *Science* essay entitled "The Tragedy of the Commons," it simply assures us that no one will take ultimate responsibility for our actions.

A sober word of warning must be added to this call for free-market environmentalism. Simply put, there are some environmental values and goods that are priceless or impossible to quantify. How many dollars is the Grand Canyon worth? How much is fruit free of carcinogens worth? How can we place a value on well water free from contaminants? By attempting to quantify items of this sort, in essence by focusing too heavily on externalities, we do run the risk of giving the impression that all cost-benefit analyses, and all public environmental decisions, can be devolved to the least common denominator of economics. Nonetheless, as large a risk as this is, it is one well worth taking given how much information we currently ignore when we set policy.

Most environmentalists, as well as those economists willing to think along these lines, recognize that we cannot immediately move from a system that ignores externalities to one that fully factors them into prices. As the examples cited above indicate, the long-ignored externalities are so large that prices would rise in an unhealthy and socially unacceptable fashion. (President Clinton's initial attempt to reform the federal grazing pricing structure failed to change the status quo precisely because the federal subsidies have been so huge. Had the subsidies been more modest, it would have been politically more palatable to internalize the externalities and ask western ranchers to shoulder a fair portion of the burden for grazing their cattle. After his economic reform package was passed in August 1993, President Clinton, rather than relying on the Senate to address the controversy, issued an executive order slowly raising federal grazing charges. Even so, his plan to increase prices from $1.86 to $4.28 per cow and calf over three years still leaves federal prices more than three times the going rates on private lands.) That we have a long way to go does not mean that we should postpone this important journey, however. Society will be significantly better off when we recognize that the pollution of our air and our drinking water and the destruction of our forests, grasslands, and coral reefs come with hefty price tags.

At the local level where such externalities have already been incorporated, some dramatic successes have been realized. The best example is the situation that has arisen when communities have acted responsibly in the face of limited landfill space. A number of communities facing this dilemma have gone from a flat fee for trash removal paid through property taxes to a situation in which residents pay varying amounts depending on the volume of trash they generate. With a flat fee there is no incentive either to recycle or to minimize the total volume of trash produced. With a pay-as-you-go policy in place, communities have created the appropriate incentives and have dramatically reduced their trash-hauling costs as well as the rate at which limited landfill space is being used up.

Of course, creative thinking about externalities and federal subsidies is not particularly "liberal." Although even the suggestion may seem like rewritten history, former secretary of the interior James Watt has left an environmental legacy in this one regard of which even the staunchest environmentalists can be proud. Yes, this is the same James Watt who divided the country into Americans and liberals, who likened environmentalists to Nazis, who systematically removed from decision-making power in the Department of the Interior anyone who had any ties to environmental groups, and who claimed, "We can learn a great deal from Walt Disney's crowd management principles." Although as secretary he relaxed strip-mining controls, voluntarily halted the acquisition of new land for addition to the national park system, and opened more federal land for oil, gas, and coal exploration than any previous interior secretary, he was also a strong supporter of the 1982 Coastal Barrier Resources Act. Known as CoBRA, this act designated almost half a million acres of barrier reefs and islands along the eastern seaboard and Gulf coast as being worthy of protection. The government designated these acres, among the most biologically productive and environmentally precarious in the nation, as subsidy-free zones, refusing to provide any subsidies for development projects of any sort. The range of federal subsidies that were cut off is impressive: FHA-guaranteed loans, grants for highways, bridges, and ports, and urban development grants, for example. Additionally, the government has refused to underwrite flood insurance or offer disaster relief for new projects in these hurricane-battered environments.

The establishment of subsidy-free zones was compelling, even to the environmentally unenlightened Reagan administration, for a number of reasons. First, by reducing development on these barrier islands, it protected life and property. Second, it yielded a significant monetary saving. Supporters claimed that $5.5 billion to $11 billion would be saved by the government over the first twenty years. Third, CoBRA required virtually no growth in the federal bureaucracy. Frank McGilvrey of the Department of the Interior said it best: "Although 13 federal agencies must withhold subsidies under the act, only one additional full-time employee has been required: me!" Finally, CoBRA was consistent with the Reagan administration's feelings about deregulation. CoBRA did not prevent private individuals from using their land as they saw fit, it merely stated that the government would not subsidize those activities determined not to be in the best interest of the public.

Since 1982, the CoBRA concept of subsidy-free zones has found widespread favor with legislators, if not with land developers. In 1986, for example, Congress passed the Colorado River Floodway Protection Act denying all federal subsidies to development on the floodplain of the lower Colorado below Hoover Dam. That same year Congress also approved the "swampbuster" and "sodbuster" provisions of the Food Security Act. These provisions lift all federal agricultural subsidies when agricultural crops are planted on wetlands or highly erodable lands. Each of these actions, in addition to saving millions of federal dollars, is helping to preserve unique ecological habitats.

Land developers have expressed an alternative view. At public hearings in Texas, developers accustomed to thinking of their subsidies as entitlements challenged CoBRA as a "taking" of their private property rights. The developers audaciously contended that to deprive them of subsidies would deprive them of the free exercise of their constitutional rights as well. The Department of the Interior quickly dismissed this bizarre logic: CoBRA allows developers to do whatever they wish with their land, but the federal government will not subsidize those actions that undermine a congressionally mandated public good.

With a bit of creativity, the subsidy-free zone concept might become a very powerful environmental tool. For example, one of the greatest threats to our national parks is development just outside their borders. Road construction, mining, lumbering, grazing, and housing developments—all currently subsidized by the federal government—isolate parks and reduce the amount of wildlife they can support while destroying the scenic vistas for which our parks are so highly prized. If we want to protect these areas without buying more land or regulating private behavior, subsidy-free zones could be the answer, or at least a very effective, low-cost first step.

What is obvious is that if we are to successfully weather the environmental assault we are leveling against our planet, we are going to have to change our ways of thinking. One of the most important people prodding us to think more creatively and more benignly is environmentalist-physicist Amory Lovins. MacArthur Fellowship–winner Lovins, founder of the Rocky Mountain Institute, an environmental think tank, has an enviable record of looking freshly and imaginatively at seemingly intractable environmental problems. Jon Louma, writing in the *New York Times* (20 Apr. 1993) has quoted John S. Hoffman, the individual responsible for energy efficiency programs at the EPA, extolling the virtues of Amory Lovins: "I think it's fair to say he truly has played the role of a prophet. A lot of what's happened in this field simply wouldn't have happened without him. At this point, everyone looks to Amory to get new ideas."

Such has not always been the case. After Lovins published what is probably his most important paper, a 1976 treatise in the journal *Foreign Affairs* in which he claimed that the country could comfortably move away from both fossil fuels and nuclear power by focusing on energy conservation, he was soundly castigated on many fronts. The energy industry submitted almost three thousand pages of testimony to a Senate subcommittee in an attempt to discredit Lovins' ideas, and one opposition scientist went so far as to warn, "Should this siren philosophy be heard and believed, we can perceive the onset of a New Dark Age" (*New York Times*, 20 Apr. 1993).

Lovins inspired the wrath of the entire energy industry as well as a slew of economists because he dared to posit that the economy could continue to grow at a healthy rate without a concomitant growth of energy expenditure. He argued that we could easily satisfy all of our new energy needs by making up the difference through energy conservation. He offered the amazingly simple view that saved energy was every bit as valuable as newly created energy, while being much cheaper and environmentally less damaging. Going even further, he claimed that we could secure significant energy savings while improving our standard of living. He thus countered the prevailing sentiment that energy conservation meant enduring hardship or, as critics said, freezing in the dark. If Lovins' 1976 article was correct, it would hardly have heralded the onset of a "New Dark Age."

In fact, Lovins' estimates in that early paper *were* incorrect—they were far too conservative. In keeping with advances in energy efficiency technology, Lovins has raised his original prediction that the country could cut its energy use by 30 percent without any loss in comfort or productivity. He now claims that we could easily slice 75 percent off our energy bills without any negative effects. To help grab people's attention and reshape their thinking, Lovins has coined the terms *negawatt* and *negabarrel* and uses them in opposition to watts of energy and barrels of oil. Each negawatt is equivalent to the energy conservation necessary to reduce energy usage by one watt. Each negabarrel precludes the need for one barrel of oil. Most important, whereas a barrel of oil is a nonrenewable resource (once it's burned, it's gone), a negabarrel goes on saving energy forever.

Society, Lovins claims, should be much more interested in negawatts and negabarrels than in watts and barrels because, though they accomplish the same thing—they provide energy services to consumers—the former are cheaper and have no environmental drawbacks. In his characteristically blunt style, Lovins asserts that consumers "don't want to buy kilowatt hours. They want motor power and hot showers and cold beer" (*New York Times*, 20 Apr. 1993). And if they can get what they want for less money and less environmental impact, so much the better. For each negawatt produced (or watt saved), excess capacity is freed up to meet electrical demand at some later point. What this means is that if we produce enough negawatts, we need not produce quite so many watts. In other words, we can scale back the rate of production at existing power plants and, even more significant, forgo the production of new ones. Because power plants are so expensive to produce, a growing number of electrical utility companies have finally jumped aboard the conservation bandwagon to become the leaders in promoting energy efficiency. (Some time ago, a representative of Wisconsin Public Service, my public utility, paid a visit to my house and provided a number of energy conservation devices. He wrapped our hot water heater with an insulating blanket, wrapped some hot water pipes, looked for places to put in compact fluorescent lighting, checked for low-flow shower heads, and cleaned the coils of our refrigerator. He saved us some money while helping the utility company cut back on its peak demand—a winning situation for everyone!) In an ironic but very sound environmental turnabout, Lovins, who was viewed by electric utility executives as a pariah a few short years ago, now serves as a high-priced consultant to that same industry.

What is perhaps even more ironic is that the term "New Dark Age" should be applied when one of the most productive ways to generate negawatts is through more efficient and higher-quality lighting. And although Lovins' has been the loudest, most consistent voice on this topic, he is not alone. Large numbers of independent scientists have conducted research leading to the same conclusion. A 1990 report entitled "Conservation Potential of Compact Fluorescent Lamps in India and Brazil" written by Ashok Gadgil of the Lawrence Berkeley Laboratory and Gilberto De Martino of the State University of Campinas in Brazil demonstrates the enormous ramifications arising from changing light bulbs. Gadgil and De Martino pointed out that technology has advanced to the point where compact fluorescent bulbs, bulbs that can be screwed into any fixture currently taking a standard incandescent light, can be switched on without warm-up time and produce the same type of "warm" light as that created by incandescents. The difference is that these bulbs last ten times longer than do standard bulbs while using, at most, only one-quarter of the energy.

What difference can a few light bulbs make? According to Gadgil and De Martino, if, over the next decade, India replaced a mere 20 percent of the incandescent bulbs currently in use, it could refrain from building 8,000 megawatts of electrical capacity that it currently has on the drawing board. Estimates of the monetary savings are $430 million annually. Savings of this sort are critical because the extremely beleaguered Indian economy is being asked to support the building of power plants and electrical transmission systems estimated to cost approximately $205 billion over the next decade. Since a large amount of this money is going to have to come from international loans, a construction program of this magnitude will have devastating effects on any semblance of balance of trade in India.

Gadgil and De Martino go a step further in describing the good that might come from a conversion to fluorescent bulbs. They point out that dramatic savings could be realized if India invested in the construction of compact fluorescent bulb factories. Their calculations indicate that a single $7 million bulb factory would conserve 1,587 megawatts of electricity (or generate an ongoing 1,587 nega-megawatts of electricity). The $7 million investment, by obviating the need to construct new power plants, thus translates into a $2.4 billion savings ($950 million of which would be in foreign exchange). Investment in fluorescent light bulb factories is 320 times cheaper than investment in facilities that would power conventional incandescent lights. Already

China has recognized the economics of compact fluorescent lights. It is in the process of increasing its production of these bulbs from 12 million in 1990 to between 80 and 100 million by 1995.

Unfortunately, the United States government seems not to have the vision to facilitate the conversion to compact fluorescents. An exception to this broad, negative charge has to be made for a number of individuals at the EPA who initiated the GreenLights Program. GreenLights is a voluntary program designed to help business make the conversion to compact fluorescents. It has been more successful than anyone originally dreamed possible. Another governmental agency, however, stands in sharp contrast to the EPA's GreenLights. Rather than facilitating conversion to compact fluorescents, the U.S. Overseas Private Investment Corporation (OPIC), an agency that provides loans and guarantees for U.S. investments in developing countries, recently elected to underwrite a $150 million General Electric project to renovate thirteen incandescent light bulb factories in Hungary. According to calculations performed by Nicholas Lenssen of the Worldwatch Institute, had OPIC invested the same amount of money to build new fluorescent bulb factories instead, $10 billion in new construction costs for power-generating facilities could be avoided.

All of the foregoing has been couched in terms of economic gain. But consider the massive environmental gain that would stem directly from a switch to compact fluorescents. With significantly less oil and coal being burned to power compact fluorescents than to power the incandescents, millions of tons of airborne pollutants would not be released. And significantly less carbon dioxide, the gas most responsible for our global warming problems, would be released as well.

Now that many of the nation's utility companies are beginning to push compact fluorescent lighting on the public as a way of cutting costs, we are beginning to make progress. Nonetheless, it is still exceedingly difficult to find compact fluorescent bulbs in the average convenience store. And although a huge variety of bulbs are made, many sizes seem virtually impossible to find. Too many shop owners refuse to carry compact fluorescents because of their relatively steep price tag. What storekeepers and consumers alike need to be made aware of is that the bulbs will typically pay for themselves within a year—and turn a very nice environmental profit from the very beginning.

On a global scale, we need to convince the powers that be that energy efficiency is of the utmost importance, and cost effective, in the developing world, where population and energy demand is growing the fastest. A report released in 1992 by Congress's Office of Technology Assessment (OTA) entitled "Fueling Development: Energy Technologies for Developing Countries" puts to rest the conventional wisdom that poor countries cannot afford the up-front investment in state-of-the-art energy-saving technologies. Even though everyone seems to recognize that such technologies will pay for themselves in the long run, the argument has long been that the short-term cost was simply too large. The OTA report stood that logic on its head by demonstrating that, like the situation with compact fluorescent light bulbs in India and Hungary, investments in energy efficiency will more than pay for themselves immediately by dramatically cutting the need for construction of new power plants. The report went further and called on the United States and the World Bank to begin to emphasize energy conservation in their lending policies rather than focusing so heavily on promoting the construction of power plants.

Lovins is quick to remind us that there are many other ways to save energy while improving performance here at home. "In the lower 48 states are two supergiant oilfields, each bigger than the largest in Saudi Arabia; each able to produce sustainably (not just to extract once) almost 5 million barrels per day for less than seven dollars a barrel; each capable of *eliminating* U.S. oil imports before a new synfuel plant or power plant or frontier oilfield could produce any energy whatever, and at about a tenth of its costs. They are the 'accelerated-scrappage-of-gas-guzzlers oilfield' under Detroit and the 'weatherization oilfield' in the nation's attics." (As bright as Lovins is, he isn't necessarily the clearest writer. His wife, attorney Hunter Lovins, who is head of the Rocky Mountain Institute, has been quoted in reference to his 1976 paper: "I knew right away it was the most significant work I'd seen, even though he wrote it in Martian" (*New York Times,* 20 Apr. 1993). She has described part of her present job as translating from Lovins' Martian to readable English.)

The government has been fighting with Detroit for ages to improve the fuel economy of the cars it designs. In 1975, when the automobile industry seemed incapable of making any progress on its own, Congress passed the first federal standards mandating fuel efficiency. With automobile exhaust causing some of our most serious air pollution problems, from acid rain to smog, we all have a vested interest in seeing that car use is minimized and that better-quality vehicles are produced. Unfortunately, industry and some governmental officials have conspired to ensure that little progress is made.

In 1991, when Senator Richard H. Bryan (D-Nev.) introduced a bill entitled "The Vehicle Fuel Efficiency Act," he was attacked by industry flak as well as by officials in the federal Department of Transportation, who stirred up quite a tempest when they "renamed" Bryan's bill the "Highway Fatality Act." Their reasoning, although terribly flawed, was straightforward: the only way to make a more fuel-efficient car is to make a lighter car, and since lighter cars are inherently less safe than heavier ones, fuel-efficient cars are less safe than gas-guzzlers. Similarly, an organization called Citizens for Vehicle Choice began running television commercials showing, in excruciatingly slow motion detail, what happens when a 4,000-pound luxury sedan crashes head-on with a 2,300-pound subcompact.

Passenger safety and fuel economy are not mutually exclusive goals. Nor is it necessary to wait for technological breakthroughs in order to achieve spectacular advances in fuel economy. Rather, Lovins argues provocatively that, if automobile manufacturers wanted to, cars far safer than those on the road today could regularly achieve one hundred miles per gallon. And, he asserts, they should cost no more than those currently on the market.

Lovins argues that new cars should be fabricated out of space-age composites rather than steel. The composites have a number of virtues and only one apparent drawback. Because the composites are "stronger, bouncier, and more shock-absorbing," they are significantly more crash worthy. Additionally, by using such materials, manufacturers would reduce a car's weight by about half. With every two-hundred-pound reduction in auto weight yielding approximately a 5 percent increase in fuel economy, Lovins is talking about massive savings. The drawback? Composite materials are significantly more expensive to make. But because it is possible to mold composite materials into single, complex pieces, savings in the manufacturing process more than offset the additional cost of the materials. Chrysler, for example, has found that it could make a complete car body out of five to seven snap-together composite parts rather than the four hundred separate parts currently welded together in today's average automobile.

Not everything Lovins advocates is high tech. He claims that the typical family-car air conditioner is "big enough for a sunbelt house, heavy, costly, high-maintenance, full of CFCs. and a drain on engine power." Currently air conditioners need to be this big to reduce the heat load of cars sitting in the sun all day. But Lovins says that the load can be sharply curtailed by using light-colored paint and upholstery and heat-reflecting glass.

From improving aerodynamics and braking systems to changing basic design to one in which each wheel is "driven by a small 'switched-reluctance' electric motor," Lovins believes that a wedding of present technology with unbridled imagination will yield low-cost, fuel-efficient, safe automobiles. We can have such cars, he asserts, if we, or our government, demand them.

We can also make significant progress on the other side of the transportation equation, automobile use, if we take a closer look at some of the subsidies and externalities associated with driving. As with so many other environmental issues, the Reagan and Bush administrations were dead wrong on these issues as well. During the Bush years the Department of Transportation released a report designed to set the nation's transportation agenda for the coming century. The report, entitled "Moving America: A Statement of National Transportation Policy," is impressive for its consistency if for nothing else—it effectively ignored all the environmental woes stemming from our overreliance on the automobile and came down on the wrong side of nearly every issue it addressed.

Three striking examples should suffice to demonstrate how the writers of the report failed to think creatively about using subsidies to improve our environmental standard of living. Instead of recognizing the absurdity of our use of the personal automobile for the bulk of interurban transport, the report recommended eliminating federal assistance for passenger rail service. Without such federal assistance it is highly unlikely that Amtrak will be able to meet its goal of becoming financially self-supporting in the near future. Without such assistance it is not even clear that the train system will survive.

The report also recommended further slashes in federal operating assistance for urban public transportation. The cuts endorsed are on top of a 50 percent reduction in federal funding that had occurred over the previous nine years. Without continued federal support, many cities will find themselves unable to piece together workable mass-transit systems. Air quality will continue to deteriorate, urban congestion will become even more unmanageable, and our reliance on imported fossil fuels will continue unabated.

Finally, the report made no suggestions about overhauling an archaic tax system that promotes the use of the automobile over public transportation. Federal tax laws permit employers to deduct whatever they spend on employee parking, but laws offer no such benefits to employers who contribute to employee expenses associated with either mass transit or ride sharing. With a federal budget deficit spiraling out of control, it was not surprising that the Bush administration was not favorably inclined toward additional tax breaks. However, in financial terms, trading the parking tax break for a mass-transit tax break would, at worst, be revenue neutral, while representing a huge environmental gain. Barring a trade, we would clearly have been ahead financially if the parking tax break were eliminated entirely. Such a simple action, by removing some of the incentive for urban and suburban employees to drive to work, would yield significant improvement in air quality.

There are two absolutely critical points to keep in mind about energy conservation strategies: they work, and they don't necessarily mean a reduction in quality of life. From 1973 to 1986 the U.S. economy grew by 35 percent, although, owing to modest attention to conservation and development of renewable energy sources during this period, energy consumption remained constant. Arthur Rosenfeld, a physicist at the Lawrence Berkeley Laboratory, claims that even our sporadic efforts at energy conservation during this period have produced a yearly savings of $150 billion without any decline in standard of living. Yet the average American, mostly because of profligate habits, still uses twice as much energy as does the average German, Swiss, or Japanese citizen who enjoys a comparable standard of living.

The good news is that enormous progress can be made, but only if we focus our attention on the problem of population growth. Not even Lovins will solve our problems if we continue to pretend that humans are not subject to the same laws of nature as are other species. We must come to grips with the fact that the earth has a carrying capacity for human beings.

As with energy conservation, a society based on the concept of sustainability is not one that requires sacrifices of its citizens—if it's done correctly in any case. What is required to do it correctly is a willingness to turn to nature for advice. When we have been willing to model natural processes rather than create high-tech "solutions," we have achieved some impressive successes. Consider two ways of dealing with the tons and tons of raw sewage generated every day. One standard, high-tech process involves sequentially treating sewage effluent with an array of potentially deadly chemicals; its ultimate goal is to kill all pathogenic organisms, restore an appropriate chemical balance to the inflow, and produce clear, potable water as an outflow. The succession of chemicals are necessary, in large part, to counteract the ill effects of the previous chemical. Donella Meadows, in her book entitled *The Global Citizen,* graphically describes one such sequence: "The effluent is made alkaline with sodium hydroxide and then blasted with chlorine gas. The chlorine oxidizes ammonia to nitrogen gas, which bubbles off into the atmosphere. Excess chlorine is inactivated with sulfur dioxide to produce sulfate and chloride. Then the whole business is filtered through activated carbon to remove any remaining chlorine." When all goes well, clean water flows out the end. When things go wrong, any number of carcinogenic and toxic by-products can be released. Gas masks, warning signs, and evacuation plans are all central parts of a high-tech, chlorine-based sewage treatment plant.

John Todd, executive director of Ocean Arks International, has developed a very different type of sewage treatment plant. Instead of a witch's brew of chemicals, he uses the properties inherent in organisms to perform the same functions with the underlying assumption that in balanced ecosystems there are no waste products—there is always some species capable of making a living off the by-products of another species. Building on this principle, Todd has created a greenhouse ecosystem through which raw sewage flows. With denitrifying bacteria, algal ponds, shrimp, bass, trout, snails, watercress, irises, and marigolds naturally performing the same functions as chlorine, sodium hydroxide, ammonia, and sulfur dioxide, a completely safe, elegant system is created. As Donella Meadows writes, "the worst imaginable mistake might kill off some snails, but it won't require an evacuation." Added benefits are that Todd's sewage treatment plants are aesthetically attractive greenhouses rather than the standard foul-smelling facilities and they provide commercial opportunities such as raising flowers and aquarium plants and animals.

To date, Todd's successes have been modest given the difficulty convincing municipalities to approach an age-old problem from a new perspective. To my mind, one of the most exciting projects Todd has completed is a wastewater treatment facility for the Boyne River School, run by the board of education of the city of Toronto. Upon entering the school building, visitors are confronted with a fountain of cascading fresh water. That water, purified by a self-contained series of four linked ecosystems, is a testimony to students, teachers, and visitors of how it is possible to live gracefully and harmoniously with nature without making any sacrifices in quality of life.

Again, looking to nature for advice and thinking about sustainability lead to creative problem solving. Currently we spend more than $130 billion annually on pesticides in our misguided commitment to high-tech farming practices. The externalities, contaminated air and water, industrial accidents such as that occurring at Bhopal, poisoned workers, tainted fruits and vegetables, disturbed ecosystems, and hazardous waste dumps, are every bit as daunting as the direct cost of the chemicals. Given, as we saw in Chapter 8, that we continue to lose the same percentage of our crops to pests as we did at the turn of the century, what benefits have we gained? It is time to rethink our reliance on such chemicals and turn to biological means of pest control. The field of integrated pest management, relying on so much more than merely spraying insects and weeds with chemicals, has been shown to be economically profitable and environmentally sound wherever it has been given the opportunity. Like John Todd's sewage treatment plants, integrated pest management needs to be given more opportunities.

We know from information taught in every basic ecology course that wholesale insecticide use cannot possibly succeed. Like broad-spectrum, oral antibiotics, widespread use of common nonspecific insecticides creates serious problems because of their basic "side effect": both good and bad insects are killed. If the bad insects are the ones that eat our crops and the good insects are the ones that eat the bad ones, basic population biology tells us that our problems will only get worse as we continue to spray. In ecological terms, the good insects are predators preying on the bad insects (the prey). Spraying will equally decimate populations of both insects, but the bad ones will be able to rebound much more quickly. The reason is relatively straightforward. In predator/prey systems, predators control the death rate of their prey (by eating them), and, in turn, the prey control the birthrate of their predators (by providing nourishment). When population numbers of both decline, therefore, the prey's death rate is dramatically reduced (because there are many fewer predators) and the predator's birthrate is dramatically reduced (because there is little food). The birthrate of the prey is much less impacted, and thus those insects feeding on crops are capable of rebounding first. Then, finding themselves initially in an environment largely devoid of predators, they are capable of reaching higher numbers than before. So insecticides exacerbate exactly the problems they were ostensibly designed to solve. If we turn to natural predators rather than to chemicals, a technique advocated by proponents of integrated pest management, we can create a stable cycle of control without many of the external costs and consequences of pesticide use.

We have to learn to modify our worldview to accept the low levels of crop loss to pests which come with stable cycles of control. But constant low-level losses are clearly better than the massive losses that result when outbreaks occur. One example will demonstrate just how immense the consequences of large outbreaks can be. In 1988, in a plague of biblical proportions, billions of voracious desert locusts were once again eating their way across a food-starved Africa. These swarms of locusts, each occasionally encompassing 360 square miles with as many as 200 million insects per square mile, quickly decimated meager grain crops recovering from the extended drought that plagued the continent. The Food and Agriculture Organization (FAO) of the United Nations, the international body most active in fighting the locust swarms, devoted most of the money at its disposal to a program of spraying with the pesticide dieldrin because it was cheap and relatively easy to apply, even though dieldrin was banned in the United States, as well as in most other developed countries, because it had been shown to cause, in humans, serious kidney damage, tremors, convulsions, respiratory failure, and central nervous system depression as well as cancerous tumors. Equally important, dieldrin is extremely stable and able to persist in the environment for many years, allowing easy, massive accumulation in wildlife. Indeed, tests have found birds and insects containing concentrations of dieldrin as much as 120 times greater than the amount found in soils. Finally, and perhaps most frightening, insects have repeatedly demonstrated the ability to develop dieldrin resistance. Dieldrin, after years of use, is no longer effective against boll weevils, horseflies, and mosquitoes. Furthermore, studies have suggested that once resistance to dieldrin has developed, resistance to other pesticides evolves quite rapidly.

With the locust crisis upon it, and faced with other, more expensive and not necessarily effective alternatives, the FAO opted to ignore the long-term consequences of dieldrin use. Given that failure to act might well have led to the starvation of hundreds of thousands of additional people, the FAO's decision cannot be criticized very strongly. The dilemma that FAO found itself in, however, is exactly the type that we must work to avoid in the future. Rather than spending all of our limited resources combating immediate environmental crises, we should devote a significant portion to basic research before the problems get out of control. A shockingly small amount of money, for example, has been spent to investigate programs that might have less immediate but greater long-lasting effects than heavy insecticide spraying during outbreaks. More effort, for example, could be put into a search for bacterial and fungal pathogens of the desert locust to serve as biological control agents. The use of such agents would create a stable cycle that would preclude the periodic outbreaks that have caused so much destruction. In this regard it is worth noting that the red locust outbreaks in Mauritius were successfully controlled by the importation of Indian mynah birds in 1762.

Ken Snyder of the National Audubon Society has said in a phone conversation, "To respond to the problem only when it is at hand, Africa will find itself at the mercy of large cyclical outbreaks requiring large quantities of pesticides. A preventative program which integrates early warning technology, biological controls and chemical pesticides is ecologically and economically a better solution to this recurring problem."

The locust problem is representative of many of our biggest ecological quandaries. All are easier to deal with earlier rather than later, and often biological rather than technological solutions will work best. As Henry David Thoreau said so long ago, "what is the use of a house if you don't have a decent planet to put it on?" We have an enormous distance to go to correct some of the critical environmental mistakes we have already made and to keep from making any number of new ones. With an alteration in perspective, with acknowledgment that growth is not always best, and with an increased appreciation and understanding of the wonders of the biological world, such a change is possible. But as the ozone hole enlarges, as global temperatures rise, as species become extinct, as old-growth temperate forests and tropical rain forests are decimated, as groundwater becomes increasingly polluted, and as air becomes increasingly fouled, our time is running out.

Although we might be hard pressed not to be pessimistic about our environmental prospects, some optimism is clearly in order. An environmental ethic has increasingly begun to permeate the elementary school curriculum in many communities, and, as the Jesuits have stated so clearly, most lifelong beliefs are formed very early. When our children's awareness is combined with an appreciation for the scientific method and a healthy dose of skepticism, our schools might yet produce a generation of scientifically enlightened individuals.

In the meantime, there are other reasons to be mildly hopeful. Although creationists continue to play a disconcertingly large role on far too many local school boards, and although there is little evidence that they have moderated their stance on science education in general or on environmentalism in particular, there is evidence that the mainstream has moved quite a distance from the creationist's extreme positions. It seems unlikely in these waning days of the twentieth century that a secretary of the interior could get away with some of the statements made more than a decade ago by James Watt. It was Watt, after all, who claimed that humans should not worry about preserving natural resources because Armageddon and the Day of Judgment were fast approaching; after that day none of us would have any further need for natural resources. Indeed, one could argue that Watt's position was just an extreme version of what historian Lynn White Jr. argued in 1967 in his seminal *Science* magazine essay entitled "The Historical Roots of Our Environmental Crisis" (10 Mar. 1967). White argued persuasively that Christianity "not only established a dualism of man and nature but also insisted that it is God's will that man exploit nature for his proper ends." In recent years a growing number of theologians and religious studies scholars have attempted a more positive integration of environmentalism and mainstream Christianity. An example of how far we have come in this regard can be seen in the work of Sallie McFague of Vanderbilt Divinity School. McFague's latest book, *The Body of God: An Ecological Theology* (1993), argues that the universe itself is an embodiment of God and thus is deserving of our respect and our care. In contradistinction to White's view, McFague maintains that "if the stereotype, even if it is just a stereotype, that we are the dominant creatures and everything is there for our use is connected in any sense with a religious view, then that's part of the problem."

Similarly, the United Church of Christ's Commission on Racial Justice gave the entire field of environmental racism a huge boost with its publication in 1987 of its report entitled "Toxic Wastes and Race in the United States." The report's major conclusions were as breathtakingly simple as they were unsettling. "The . . . study suggests that the disproportionate numbers of racial and ethnic persons residing in communities with commercial hazardous waste facilities [are] ... a consistent pattern. Statistical associations between race and the location of these facilities were stronger than any other association tested. The probability that this occurred by chance is less than 1 in 10,000." More accurately, the racial composition of a community is the single best predictor of the presence of commercial hazardous waste facilities. Although, not surprisingly, toxic waste sites are more prevalent in lower rather than in higher socioeconomic communities, the economic status of residents was not nearly as good a predictor of toxic waste sites as was race itself.

Again, none of this is to say that the creationists and other representatives of the religious right have become receptive to a fresh environmental paradigm. That is demonstrably not the case. Some publications of these groups have called environmentalism the "new satanism" and have asked, "Is the ban on wetlands a return to paganism?" As environmentalism becomes better understood, extreme views of this sort should become increasingly out of fashion.

Index